C0-ATR-923

About Island Press

Island Press, a nonprofit organization, publishes, markets, and distributes the most advanced thinking on the conservation of our natural resources—books about soil, land, water, forests, wildlife, and hazardous and toxic wastes. These books are practical tools used by public officials, business and industry leaders, natural resource managers, and concerned citizens working to solve both local and global resource problems.

Founded in 1978, Island Press reorganized in 1984 to meet the increasing demand for substantive books on all resource-related issues. Island Press publishes and distributes under its own imprint and offers these services to other nonprofit organizations.

Support for Island Press is provided by Apple Computer, Inc., Geraldine R. Dodge Foundation, The Energy Foundation, The Charles Engelhard Foundation, The Ford Foundation, Glen Eagles Foundation, The George Gund Foundation, William and Flora Hewlett Foundation, The Joyce Foundation, The John D. and Catherine T. MacArthur Foundation, The Andrew W. Mellon Foundation, The Joyce Mertz-Gilmore Foundation, The New-Land Foundation, The J. N. Pew, Jr. Charitable Trust, Alida Rockefeller, The Rockefeller Brothers Fund, The Rockefeller Foundation, The Florence and John Schumann Foundation, The Tides Foundation, and individual donors.

Time *for* Change

Time *for* Change

A New Approach to Environment and Development

HAL KANE
edited by LINDA STARKE

Prepared for the U.S. Citizens Network on UNCED

ISLAND PRESS
Washington, D.C.• Covelo, California

Library of Congress Cataloging-in-Publication Data

Kane, Hal.
Time for change: a new approach to environment and development/ Hal Kane with Linda Starke.
p. cm.
Includes bibliographical references and index.
ISBN 1-55963-156-2.—ISBN 1-55963-155-4 (pbk.)
1. Economic development—Environmental aspects. I. Starke, Linda. II. Title.
HD75.6.K36 1992
363.7—dc20 91-41029
CIP

Printed on recycled, acid-free paper

Manufactured in the United States of America

10 9 8 7 6 5 4 3 2 1

Contents

Preface

The U.S. Citizens Network on the United Nations Conference on Environment and Development (UNCED) grew out of a meeting of some 200 representatives of U.S. non-governmental organizations (NGOs) held in Washington, D.C., in October 1990. With less than two years left before the U.N. meeting in Rio, which many felt to be the most important gathering of its kind in our time, it was clear that the NGO community could benefit from sharing its views on the many vital issues to be addressed at the Earth Summit, and from presenting these views to the government and as broad a range of people as possible. The Rio meeting has been the focus of a great deal of interest in much of the world, but the American public knows little of its import yet. In part, this is due to the little substantive attention it has received from the U.S. government and its many agencies, all of which should be concerned about their impact on our environment and development future.

Since its enthusiastic launch in late 1990, the Network has worked with hundreds of groups around the United States to bring the Rio meeting to the public's attention and to let the U.S. government know the full range of views on issues to be addressed there. It has participated in the five Round Tables held by the U.S. Council on Environmental Quality to discuss the draft of the government's report to the UNCED. More important, perhaps, it has brought together groups from around the country who did not realize how much they had in common, or how much they should. In that sense it is truly a network, not a lobbying group on single topics or just a discussion group on the U.S. government's report.

Cooperation and learning about each other's viewpoints is what the Network is all about. That work will continue through the Earth Summit in June 1992, and will no doubt forge new alliances to ensure that sustainable development—the rallying cry of groups around the world—remains at the top of the U.S. agenda.

Time for Change attempts to point out the connections between the many different strands that must create sustainable development. The division of our thinking, and our institutions, into discrete categories called energy, industry, population growth, transportation, women's issues, housing, health care, and so on is one of the major impediments to a sustainable future. Enabling people to understand the intimate links between and among these seemingly unrelated categories of issues is a difficult task—but a necessary one nonetheless. Part I of *Time for Change* looks at policy options for the future at the national and international levels. Part II considers the evidence that our environmental and social systems are under growing stress.

The Citizens Network includes dozens of organizations and hundreds of individuals, and this book does not try to cover their wide-ranging views on the topics it addresses. But we endorse fully the principle on which it is based—the all-important connections among our economic, environmental, social, and cultural systems.

Readers who wish more detailed information on the many topics that can only be touched on in this volume should consult several valuable publications that are now fixtures in these fields, such as the *Global Ecology Handbook* put out in 1990 by the Global Tomorrow Coalition; the *Human Development Report* published annually by the U.N. Development Programme; UNICEF's annual *State of the World's Children;* the *World Development Report* that comes out each year from the World Bank; the annual *World Military and Social Expenditures* by Ruth Leger Sivard; *World Resources,* published biannually by the World Resources Institute; and the annual *State of the World,* available from Worldwatch Institute.

We hope that *Time for Change* provides a focus for people seeking new alliances to take effective action based on information increasingly available on the burdens being placed on our lives and on our home, the earth—burdens that are jeopardizing our common future.

Frances Spivy-Weber
Chair, Administrative Committee
U.S. Citizens Network

Washington, D.C.
September 30, 1991

Acknowledgments

Putting together a book that by its very nature covers a broad range of topics in discrete disciplines that too seldom interact requires the cooperation and commitment of a great many people. *Time for Change* benefited from just such cooperation and commitment. The principal architects of the book's concept were the members of the Citizens Network's Synthesis Committee—Larry Martin of The Other Economic Summit, Americas in Washington, D.C.; Bill Pace of the Center for Development of International Law in Washington, D.C.; and George Paul of the Foundation for BioDiversity in Phoenix, Arizona—working with Network Administrative Committee Chair Frances Spivy-Weber of the National Audubon Society's Washington office.

A special thanks to Commonweal and Friends of the Earth (FOE) for permission to use the quotes scattered throughout this book that are drawn from *Earth Summit: Conversations with Architects of an Ecologically Sustainable Future* by Steve Lerner, published in 1991 and available through either Commonweal in Bolinas, California, or FOE in Washington, D.C.

The 19 Working Groups of the Citizens Network provided advice on what the book should include. Although it is not possible to list here the many members of those groups, their contributions are gratefully acknowledged. They continue to study the issues that will be debated at the U.N. Conference on Environment and Development in June 1992, as well as those that should be, but that are not, on the agenda. Position papers from the following groups were particularly valuable: Energy, Jack Gleason (Convener); Ethics, Development and Environment, Peter Adriance, Angela Harkavy; Biodiversity, George Paul; Sustainable Agriculture, Donna Batcho; Economics, Financial Institutions and Military Spending, Larry

Martin; Oceans, Tiffin Shewmake; Freshwater, Marda Mayo; Urban Environment, Stuart Chaitkin; Global Climate Change, Navroz Dubash; Forests, Ken Snyder; International Institutions, Bill Pace, Kathy Sessions; Trade, Neil Ritchie, Kristin Dawkins; and Biotechnology, Eileen Nic. The Working Groups can be reached through the Network's office at 300 Broadway, # 39, San Francisco, CA 94133 (tel: 415-956-6162; fax: 415-956-0241) or through the Chairs of the specific group.

Time *for* Change

CHAPTER 1

Cycles & Connections

Consider an ecosystem. It is a concept from nature, but implicit in it are patterns that can be found in even the most artificial of human situations. As such, it offers a manner for considering cultural, economic, and governmental decisions.

In forest ecosystems, moisture moves in cycles, from clouds to rivers and plants and soil, and through the air to distant territories. Changes in one part affect the other parts, even though at first they might seem separate. And many nutrients used by trees are not stored in the soil, but in other trees and plants, which return them to the growth process once they die. If the trees are cut and carried away, the nutrients never return to the soil and it is left barren.

The same can be said for culture. Art evolves over the years, as the work of some artists affects others, and new movements take new directions accordingly. Music passes through phases where songs and musicians reinforce the work of other artists in a direction approximately common to all of them, and a genre evolves. In day-to-day culture, people take cues from those around them to fashion a way of life compatible with their friends and neighbors and harmonious with the place where they live.

Similarly, in economics the opportunities for corporations are affected by government, other businesses, and natural systems. Each contributes one link of the process through which products are developed. And the health of one market affects the health of others, as the demand for products of one comes from the profits made in another.

In government, the same happens. Changes in one agency affect others. Decisions intended to affect industry also affect the environment, and vice versa. Price incentives on certain sources of energy encourage or discourage research on and development of other sources. Subsidies on certain crops affect the structure of the economy, the budget, and the ways of life of people. State and local governments, and nations, operate in interconnected ways, like ecosystems.

In all these systems each part depends on the others. Problems in one part are problems for all, because a weak link in the cycle will block the flow of resources to the other parts of the cycle.

Decisions can take these cyclical, interconnected patterns into account, or they can ignore them. Examples exist of both. But in the areas where we face long-term, reoccurring problems, short-term, linear decision making is often at the root. Such decision making produces cycles of its own—downward cycles under which one problem adds to others, and where progress is difficult because it requires more than local improvements. It needs comprehensive remedies.

This can be seen in the difficulties of many governmental agencies. U.S. policy toward the environment and social and economic development provides powerful examples. No U.S. government plan exists to treat social, economic, and environmental problems in the cyclical, interconnected fashion in which they work. Rather, the agencies directed toward these goals exist separately. Those that make decisions over the processes that affect the environment act independently from those setting policy for industry, transportation, trade, energy, urban planning, and social welfare. Policies are not cycled among the agencies, but rather divided among them.

Similarly, public health is treated as one goal, environmental protection another, and food and drug safety a third. But in fact they are so closely related as to depend heavily on each other. Under this system, linear policies are patchwork attempts at cleaning up reoccurring messes, and piecemeal solutions that cannot address the completeness and broadness of environmental and public health problems.

Capabilities

Every year, human systems grow in size and scope. Medical care reaches more people, and does so with greater effect. Industry touches the environment more extensively, and creates more innovative technologies for protecting it. The population of the United States increases incrementally, and that of the world, exponentially. Consumption rises impressively, additional materials are used and discarded, and chemicals replace organic substances. Transportation gets faster and easier, and communication becomes more visual and more far-reaching.

Human creativity can be expressed in more ways than ever, and can alter more deeply the natural and societal systems it focuses on. The closely linked nature of today's systems poses a challenge under this situation. Unexpected changes occur from developments in seemingly isolated areas, as we discover connection after connection among developments previously considered separate. Human imagination and innovation are the resources with the potential to deal with such complexity.

Certain topics tie together many of the critical areas that need to be addressed today, and imagination and innovation have the most potential when they work on those topics. They include resource accounting, investment, education, urban policy, population, trade, debt, energy, demilitarization, technology, and human rights. Each is a link between areas that can seem distinct but are not. Improvements in human rights, for example, lead to improvements in many areas, such as better protection of the environment, fewer military conflicts, and increased opportunities for investment. Improvements in technology offer efficiency and adaptability to most human activities.

But many of these systems are taken for granted. People regret the loss of forests, for example, but they rarely attribute it to patterns of international trade, or to a failure of economic accounting systems to include all the functions of a forest, including the ecological ones, when valuing timber.

International trade is the system under which natural resources are allocated. By following its dictates, people avoid the need to make value-based decisions on how to use the environment. If the international market has a place in it for North American timber, then the timber will be cut. If not, it will remain. Ethics are not involved. The market formula partially replaces human decision making, and few evaluations are made to see if the way resources are used is compatible with national goals and needs.

In the 1990s, both the opportunities and the needs are greater than ever to make social, economic, environmental, and other systems more compatible. The adjustment this entails will require changes in the underlying practices like trade that affect many other areas. The pace of change in natural systems is so great that irreversible damage has already occurred, and some signs suggest that drastic change may be occurring. (See Chapter 5.) Worldwide population growth has put all systems—social, economic, cultural, environmental, and others—in a position where they must either deal with their interconnectedness or reduce human potential and options.

The 1990s will be the decade when society's institutions must work to solve the problems of other institutions as well as their own, and when countries have to solve the problems of their neighbors as well as their own. In many cases, the best opportunities for improvement do not lie with the people expected to make the changes; they lie with people whose activities are at the source of the developments, who have access to particular technologies, or who control a point of linkage among various activities.

Once again, the environment illustrates this interdependence. Shared rivers cross national boundaries, the atmosphere knows no limits in geographical area, and ecosystems depend on what lies on both sides of political borders. Chemicals dumped into a river in one country make their way into another. Erosion that fills river beds with topsoil in one country changes the water balance of a neighbor. Acid rain, the greenhouse effect, ozone depletion, and airborne pollution all travel to distant regions or affect the globe as a whole. The solutions to these problems do not necessarily originate in the country most adversely affected.

And ecologists who study the degradation of natural systems often are not equipped to reverse the decline. That task lies with the people who make decisions about industry, transportation, energy use, commerce, packaging, and technology. Much of the best potential for environmental improvements rests in the hands of organizations that traditionally have not considered environmental goals as one of their top concerns, and that have not educated their staffs on the environmental roles they could play. Problems continue to be treated in isolation, with little cooperation among the various institutions and individuals whose actions affect more than their own areas of expertise.

The United States has developed specialized government and non-governmental institutions to focus on such topics as transportation, environment, legal justice, and others. It now has an opportunity to transform the mandates of those institutions into a system under which the interconnected roles of actions are turned to our advantage rather than ignored.

The U.S. Department of Transportation, for example, can reduce significantly the energy consumption of Americans and the pollution they produce by advocating a move to the most energy-efficient modes of transportation. In doing so, it would achieve a major goal that eludes the institutions dedicated to environmental protection.

The same is true of social systems. Police departments work to reduce crime, but the greatest potential for reducing crime may rest with inner-city job-creation programs or educational programs that reach the areas with the highest crime rates. In either case, it is not the people charged with crime control who sit in the best long-term position to accomplish their goals.

Discussion and communication among the various organizations that affect any interconnected element of the United States today—whether it be pollution prevention, economic development, political participation, or any other need—have the ability to improve our capacity to handle challenges. It is an ability that can augment the success we have enjoyed so far with institutions already in place and can unleash their potential to address the many problems still to be solved.

The United States in the World

The terms "developing countries" and "developed countries" are based on an erroneous perception that Northern countries like the United States have already accomplished much of what must be done to reach a desired level of well-being. Yet the United States, which consumes a disproportionate amount of the world's energy and produces an unprecedented level of waste, has a long way to go to improve the sustainability of its actions. It is consuming at a speed that pays little attention to the needs and aspirations of future generations of Americans, or to those of current and future generations of people around the world. The disproportionate responsibility of the United States and other industrial nations to change the way the world does business has been rightly pointed out by many people—and with legitimate indignation by those in what we call developing countries.

A great deal of momentum has now amassed around the concept of sustainability, and pressures have built up in areas as diverse as grass-roots social movements and the United Nations to make sustainability a priority for upcoming decisions in international institutions and fora. The United States has traditionally acted as a leader in international affairs and diplomacy. But regarding environmental issues and sustainable development, it has lagged behind.

The United States has limited its response to sustainable development to occasional rhetoric with no concrete actions, or even any evaluations of its policies to examine their roles in affecting the sustainability of development in the United States. On environmental matters, American diplomatic efforts have rarely forged forward. Instead they have held back European countries eager to reach consensus on the issues that cross national boundaries, most notably the emission of greenhouse gases into the atmosphere, the transboundary movement of acidic precipitation, and the regulation of international waterways.

At the moment, the American position of leadership in the world does not seem to include working to ensure that future generations inherit potential high standards of living, or providing direction to other nations that will allow solutions to the global problems of environment, population, and health. U.S. policy remains dedicated to growth-oriented policies and is reluctant to explore the benefits of qualitative change without growth. These missed opportunities will cost the United States dearly, both now and in the future.

They come at a time when U.S. tax and economic incentive structures can encourage or ignore the ability of Americans to develop new, healthy responses to the stresses in our social and environmental systems. Re-evaluating the primacy of economic goals and asserting the importance of broad social, community, and family values is critical in rethinking our responses to stress. Increasingly, many see the overbearing reliance on economic growth as a major contributor to the social and environmental crises we face.

Incentives to develop and promote new, cleaner technologies can also contribute to sustainable development. Sciences and technologies are now available to solve even many of the most difficult agricultural, urban, transportation, energy, and other problems, and it is the United States that has the lead in many of those areas. It should make the most of the opportunities that face it.

Moreover, it is also a time when international awareness is converging on issues of sustainability, life-style changes, the environment, interdependence, and interconnectedness. Conferences promote dialogue concerning these topics, diplomatic initiatives address them, institutions are developing programs to adapt to them, and people in many countries are learning more about them. The broad base of support is present to greet whatever leadership the United States might want to offer, both on the part of its own citizens and, in many cases, on the part of foreign governments, institutions, and citizens.

And finally, it is a time of thresholds. Environmental degradation can reach a point where recovery is not possible and where other regions and systems are pushed off balance by its severity. The cutting of a certain percentage of a forest, for example, will cause the rest to die by disrupting the cycles of water, nutrients, and life within it. Below that threshold, trees can be cut with only localized damage to the forest; beyond it, the whole will suffer. Similarly, the human body can absorb a certain amount of a number of chemicals without negative effects on health. But if those chemicals accumulate beyond a critical point, their effects can cause illnesses that threaten overall health.

More thresholds are being approached today than ever before. Exponential population growth has quickened dramatically the pace of changes taking place on the earth. Growth in industry has brought tens of thousands of new chemicals into contact with human beings and with the environment. Increased international trade has intensified specialization and production, and increased communication has deepened the interactions of people in distant regions. The 1990s are a critical moment when introspection of the actions taken by the United States and all countries, and a historical self-consciousness, are urgently needed.

The directions of development are not inevitable. They are determined by human actions and decisions. They can move us toward greater reliance on chemicals whose effects we barely understand, or toward balanced techniques in agriculture that eliminate the need for intensive use of chemicals, and toward alternative processes in industry that substitute natural or non-toxic materials for harmful ones. They can increase reliance on oil, which pollutes air and water and is concentrated in the hands of a few countries, or they can promote renewable sources of energy like solar power that pollute less and, in many cases, never run out. The United States faces these decisions today. Yet so far it has not started a discussion, nor joined the existing ones, about how to address such decisions.

The United Nations Conference on Environment and Development

UNCED or the Earth Summit, as it has been called, will be a place where decisions are made. And where the groundwork is laid for a legal and institutional framework to deal with the pressing issues of the 1990s and beyond. Preparations for it have included almost all countries and a broad range of groups within them. Indeed, both the preparations and the conference are the culmination of more than two decades of movement toward a comprehensive and coordinated international response to the challenges of environment, development, population, and interconnectedness.

In the early 1970s, a report by the Club of Rome on the *Limits to Growth* awoke many people to the limited ability of the environment to support ever increasing human populations and industrial production. The 1972 United Nations Conference on the Human Environment in Stockholm provided the first far-reaching global forum for the examination of relationships between environment and development. The creation of the U.N. Environment Programme was one concrete and welcome result of that meeting.

The ideas from that conference led to a body of literature that has increased ever since. In the United States, the President's Council on Environmental Quality (CEQ) began a project to assess the state of the global environment and the forces that might affect it, in an effort not to be caught by surprise by upcoming changes and trends. By the end of the 1970s, when the project was complete, the *Global 2000 Report to the President* notified the U.S. government of the seriousness of the changes taking place in the environment and in human society.

Political developments, however, reduced the impact of this report. The Reagan administration chose to disregard most of its information, and it reorganized CEQ's staff, reducing it in size and importance. Although CEQ is making a comeback under the Bush administration, in the early 1990s the U.S. government largely continues its policy of placing a low priority on topics of international environment and development.

In the United Nations, however, the situation is different. In 1985 the General Assembly endorsed the formation of the World Commission on Environment and Development to examine the global situation and recommend actions to respond to it. Chaired by Gro Harlem Brundtland, who became Prime Minister of Norway during its three years of work, the Commission held public hearings in eight countries and produced a report entitled *Our Common Future* in 1987. The Brundtland Commission, as it is often called, promoted the concept of sustainable development, a philosophy that provides the framework for many of the discussions of environment and development that take place today.

At UNCED, sustainable development will be the point of departure for international agreements and dialogue. A true definition of this concept requires more than a sentence to express, but the most commonly cited single definition is that in the Brundtland Report: "Sustainable development is development that meets the needs of the present without compromising the ability of future generations to meet their own needs."[1] Put in economic terms, this means living off the earth's interest without depleting its capital. This concept will be the guideline followed by negotiators in June 1992 in Rio de Janeiro, Brazil, the site of UNCED.

Whether or not the conference takes steps to increase the sustainability of current human actions and thereby increase the options available to future generations, sustainable development has provided a common language for people of diverse backgrounds. It has become a theme for people in many countries and many groups, with interests in environment, economics, population, urban structures, energy, government, agriculture, politics, sociology, indigenous peoples, and many others.

Many non-governmental organizations (NGOs) are now attempting to work toward the goal of sustainable development. They range from agricultural and health cooperatives in some of the poorest countries to well-funded non-profit environmental organizations in the wealthiest ones. Other interested NGOs include education groups, women's organizations, religious groups, students' associations, industry organizations, and financial cooperatives.

It is the North that needs development. We have the longest path to go to achieve sustainability. We in the North are much further than are people in the South from achieving a more or less self-sufficient, local, stable community.... The way we in the North can help southern people the most is to develop our countries into something more sustainable.

Gunnar Album, *Norway*
Nature and Youth
Earth Summit, pp. 82, 85

They have all participated in the UNCED process. They have educated their communities and their elected officials. In some cases, they have even been included in government delegations to UNCED preparatory conferences and to UNCED itself. In others they have been consulted by governments and by the United Nations Secretariat responsible for planning the conference. They play a formal role in the process. When the ministers gather in Rio, NGOs will be there too, to attract extensive coverage in the press, facilitate public responses to affect government actions, and to plan for the future.

The hope behind broad participation in the UNCED process is that by talking about environment and development with a broad range of people, lessons can be learned and creativity can be unleashed to respond to new challenges. UNCED offers great potential for broad-based solutions and cooperation among many countries and people. The United States will participate in UNCED but the extent to which it will devote itself, and the depth of the commitments it will make, depends on how clear a message citizens send to President Bush and to Congress. The government can choose to maximize the breadth and range of people it includes in its preparations, or it can treat the conference as a limited affair of business-as-usual politics. Its resources and potential for bringing about positive change may be the greatest of any country involved, yet its commitment is unclear.

Sustainable development at its core is about the inclusion of all the people and all the areas of expertise that touch and are touched by the natural environment and human development. The mutual reinforcement and benefit of cooperation and common understanding among people and disciplines is the greatest resource available.

Measuring Progress

At the heart of difficulties and differences among approaches to dealing with environmental and developmental problems is the problematic nature of measuring them. Sustainable development requires an integration of economics, social conditions, and environment in decision making, but economics treats environmental resources and social processes as externalities because they cannot be measured in the way that the discipline requires.

The economic systems that we use to allocate natural resources do not account for their true values. The long-term value they can offer by providing year-after-year income from sustainable activities is not included in corporate or government accounting tables. The interconnected functions that the environment provides in supporting diverse living species, regulating weather patterns and moisture, absorbing waste materials, and many others are not accounted for in the economic procedures we use to regulate our uses of natural resources.

Given this tendency not to recognize the roles that the environment plays in supporting our societies, it is not surprising that we also have a tendency to ignore the need for efforts to protect it. No simple answers exist when it comes to trying to measure the value of the functions that the environment provides, and so environmental measurements and accounting systems will continue to be an elusive point in efforts to make human actions more environmentally sustainable.

Similar difficulties plague the connections between economics and the social goals that economics is really meant to achieve. Development projects, for example, especially those in southern countries, affect cultures and societies by changing the ways of life of local people. Yet those projects have little ability, and often make no effort, to measure the cultural priorities, heritages, and needs of the people they affect. In some cases, they relocate people to new regions—one of the most culturally destructive actions in modern history. In other cases, they change daily schedules, types of housing, distribution of labor, family roles, amount of leisure time, and many other aspects of community life.

But no mechanism exists for including those changes in development planning, because no measurement exists that allows planners to compare the benefits of development with its costs. Some of the benefits can be quantified in one way or another—the number of hospital beds added, or school desks built, or kilowatts of electricity produced. But higher rates of disease as communities are exposed to new bacteria and viruses, loss of family time, breakup of neighborhoods, increased stresses, and longer times needed for commuting to work are difficult to measure, so attempts are rarely made.

Decisions are based on incomplete information. The United States may have an overoptimistic view of its state of development today, because the information it can quantify suggests that it is doing well. But the information it ignores—its possible proximity to thresholds of ecological collapse and some of the cracks in its social systems—are not included in the assessments. The failure of the United States to take up a position of international leadership on problems of environment and development may reflect the elusiveness of measuring these problems and including them in national calculations.

Much of the work to be done at UNCED and other forums on environment and development will revolve around environmental accounting and the merging of economics with non-quantifiable cultural, social, and population-related concerns. A sort of "eco-nomics" or "ecolnomics" that brings ecology and similar interconnected processes together with economics holds great potential for addressing these problems and increasing future compatibility between systems that rely on each other.

The words economics and ecology both stem from "oikos," the ancient

Greek word for house. Their separation reflects a recent development that is not consistent with their true natures. If that separation persists in American policy, it will continue to distort it. It takes no great American historian to know that a house divided against itself cannot stand.

PART I

Policy Options for the Future

CHAPTER 2

The Domestic Agenda

A great many examples exist in the United States of attempts to build an undivided house—of successful programs that approach sustainability, contribute to human resources, and include environmental and social concerns, though they rarely make headlines or get featured on the evening news. And a great deal of room exists for new programs that expand on these or find new territory for improvement.

They range from local initiatives for environmental improvements or community activities to national policies and the ways economic decisions are made. Whether the links are clear or obscure, they all reinforce each other in one way or another, and all are part of a larger effort to move toward sustainable development.

The late 1970s witnessed the beginnings of a healthy recognition on the part of the U.S. government and people of the interconnected roles played by the environment, population growth, economics, and development. The *Global 2000 Report to the President,* requested by Jimmy Carter, was at the forefront of the literature demonstrating the global nature and the urgency of what later came to be known as issues of sustainable development.

In the government, however, such efforts were shut down in the early 1980s under a different political climate. They continued in private, local, and nongovernmental areas, but by the early 1990s the U.S. government still had not reached the level of commitment it once appeared to be moving toward.

There is no lack of advice on the direction the country should be headed in and the priorities it should have. As the 1988 election approached, 18 nongovernmental organizations banded together to produce *Blueprint for the Environment,* a series of recommendations for the new president; included were pleas for a national least-cost energy plan and an official population policy that recognized the need to halt world population growth. Since then, a variety of groups have generated similar plans, demonstrating that it is possible for policy-related documents to capture the broad scope of issues contributing to sustainability.

The 1990s will be a mix of programs and policies that recognize those issues and of programs and policies that ignore them. Many of the successful ones will arise out of grassroots efforts of local activists who have discovered that building a sustainable society takes both national and local leadership and public commitment.

This chapter includes just a few examples of policies and programs that offer brighter long-term benefits and make contributions to the overall movement toward development that conserves options for the future as well as creates them for the present.

Urban Environments

A few years ago, more than 200 local New York City environmentally related groups joined together to create *Environment '89,* and prepared a thorough political plan for greening the city. With actions ranging from closing down apartment incinerators to offering low-interest loans for asbestos removal from homes and the creation of an Environmental Education Coordinator for the school system, the group tried to turn even the most squalid areas into a supportive urban environment. They focused on practical and inexpensive opportunities, and called for only a 1-percent increase in the city's park budget to fund the whole range of projects.[1]

While city politics, in the end, did not support the plan, others still await the day when the political atmosphere will be right for major steps forward in urban environment. Among them are the Green Guerrillas, who have been planting the city since 1973, and the nonprofit Bronx 2000 development corporation, founded in 1980. The people at Bronx 2000 see hundreds of millions of dollars of value in the opportunity of recycling the city's scrap paper, scrap metal, and other garbage. Such projects could be turned into an industrial development opportunity, said David Muchnick, who runs one of the firms participating in Bronx 2000, during an interview with *E Magazine.* He wants to see recycling become the mandate of a Deputy Mayor for economic development, instead of a duty of the sanitation department, which is currently struggling with it, and he would like to see recycling companies have the same access to financing that landfills and incinerators have.[2]

With more than half the world expected to be living in cities by the end of this decade, efforts to make urban areas more livable, more self-sufficient, safer, and less polluted—both visually and in their daily output of waste—are increasingly important. A new movement of urban planners, architects, designers, and city managers recently convened an international conference in Berkeley on "eco-cities". Proceedings from the conference pulled together visions from leading researchers and activists exploring new systems for "eco-cities," places where people can get around by bicycle or easily accessible public transit, where recycling is a major local industry, where energy is provided by wind power and solar technologies, where tree planting campaigns build community spirit and provide aesthetic and environmental services, and where urban gardens supply food and a focal point for local activities.[3] In a late 1991 rating of 16 cities by *E Magazine,* San Francisco and Berkeley in California and Madison, Wisconsin, came closest to being eco-cities.[4]

Education and Environment

Education supports sustainable development on many levels. Locally, it helps people understand the relationships between their personal life-styles and the condition of the environment, of their neighbors, of the local and national economies, and of people in other countries. Politically, it helps people contribute to their governments by increasing the percentage who vote, who join in community or other groups, who write articles for journals or letters to Congress, or who simply discuss their concerns with family and friends. Internationally, education improves the ability of people from different backgrounds to communicate about shared concerns in common languages, and to understand each other's needs.

Programs that focus specifically on the environment are becoming more common at all levels of schooling. In April 1990, Renew America, a Washington, D.C.-based non-profit group working toward a sustainable future, gave out National Environmental Achievement Awards to 21 programs throughout the United States that it judged best in terms of ability to protect, restore, or enhance the environment. In the category of environmental education, the Training Student Organizers Program in New York City took top honors. (See Box 2-1.)

Box 2-1: *Environmental Education*
The Council on the Environment of New York City established the Training Student Organizers Program (TSO) in 1979 as an environmental education program to train high school as well as some elementary, junior high school, and college students to organize environmental improvement projects in New York City schools and neighborhoods.

The program teaches the students about environmental issues and helps them to acquire organizing skills as they involve their peers and community residents in such projects as beach clean-ups, noise abatement campaigns, graffiti elimination, open space preservation, and paper collection for recycling. To date, student organizers have worked with more than 275 community groups and schools in the development of 250 projects serving 61 neighborhoods in New York's five boroughs.

The staff of the Council on the Environment works with classroom teachers on a weekly basis to train the students for a full school term. The students participating in TSO receive comprehensive training and academic credit while they organize environmental projects within communities and schools. The TSO staff also works with special education, homeless, and incarcerated youth. The program helps the participants to develop a service ethic and to feel more skilled, confident, and able to relate to the world around them.

SOURCE: Excerpted from Rick Piltz and Sheila Machado, *Searching for Success* (Washington, D.C.: Renew America, 1990), p. 62.

Two other programs that received special mention from Renew America are found in Maryland and Arkansas. The schools of Prince George's

County in Maryland have added environmental education to all grades from kindergarten through high school. A supervisor of Environmental Education has been appointed and liaises between the community and the schools to investigate community and educational needs for environmental understanding. An outdoor education program has been created, including a 450-acre Outdoor Education Center, where all fifth-graders spend two days and a night learning about the Chesapeake Bay, land use issues, and outdoor activities. Rather than being separate from other studies, environmental concepts are integrated into all subject areas of the curriculum.[5]

And in Perryville, Arkansas, the Heifer Project International has created a Learning and Livestock Center where youth groups, college students, and volunteers can participate in efficient and sustainable farm practices. The farm uses solar energy for its oven, food dryer, and water pump, and other renewable sources of energy. The farm's procedures demonstrate that biological cycles can be used in fertilizing crops and controlling pests, and how the use of chemicals and monocultures can destroy the natural cycles. More than 13,000 people from 47 states and 60 countries participated in this hands-on program in 1989 alone.[6]

Education like these programs, which is balanced and cognizant of the connections between economics and other areas, is basic to a country's economic democracy.

Local Self-Reliance

Education, democracy, politics, and government are all firmly linked. The success of education is a foundation for successful political participation and government suited to the goals of citizens. Although democracy has traditionally been a political concept, there can also be "economic democracy," a participation in economic systems and in their design. The success of political democracy can be undermined by a lack of economic democracy in the financial and industrial systems that are the basis of all societies.

A movement to broaden participation in this economic democracy is taking place in the United States under various headings such as local self-reliance and community economics. It seeks to capture and recycle money in communities that need it most, to develop businesses that are environmentally sound and sustainable, to reduce communities' dependence on outside sources of food and energy, to lower our dependence on the automobile by developing alternative forms of transportation, and in general to foster human and humane communities. Many of these efforts are catalogued by Susan Meeker-Lowry in her 1988 guide called *Economics as If the Earth Really Mattered.*[7]

In the area of environmental restoration, California launched a groundbreaking natural resource recovery program in 1979 called "Investing For Prosperity." Revenues from resources that were being depleted, such as oil, were reinvested in developing alternative energy source, renovating fisheries, reforesting clear-cut land, and so on. Fifty plants are now in operation that generate more than 1,000 megawatts of energy a year from wood waste. A fund was established to help small landowners be able to afford to reforest their land. The founder of that program, Huey Johnson, believes that "the most notable reason for its success is that it developed programs to renew the state's resources, not exploit fixed ones."[8]

In an attempt to keep food production local, safe, and under community control, the owners of Brookfield Farm in South Amherst, Massachusetts, began a cooperative arrangement through a nonprofit trust set up with their neighbors in 1987. Some 85 families pay hundreds of dollars for a share of the farm's harvest before crops are even planted, in part because they know that the fruits and vegetables are grown without pesticides. To reap the harvest, many shareholders volunteer to pick the produce. Although it so far touches only a tiny fraction of American agriculture, 38 small farms in 21 states have adopted this form of community-supported agriculture in the last five years. Its advocates believe it is a key to keeping small farms in business and meeting consumer demand for pesticide-free food.[9]

The Institute for Community Economics in Massachusetts tracks this and similar efforts to regain control at home. It recently reported on new community land trusts (CLTs) in Youngstown, Ohio, and Duluth, Minnesota, that are trying to deal with housing and economic problems in communities that have lost tens of thousands of blue-collar jobs.[10] CLTs were developed in the 1960s to address the issue of land tenure; they are nonprofit corporations that combine the features of private and community ownership. Land is acquired and then leased on a long-term or lifetime basis, with assistance in finding affordable financing for buildings.[11] In

Youngstown, the CLT is using funds from government Community Development Block Grants to rehabilitate abandoned homes and provide jobs and job training in the process. In Duluth, the Northern Communities Land Trust established in 1990 is working with government agencies and other community groups to purchase vacant lots for new housing and to help homeless families renovate run-down properties.

Individuals and local groups wishing to get more involved in the environmental issues of local self-reliance may want to look to the records of their neighbors as a way to pressure governments to clean up their act. The recent publication of the *1991-1992 Green Index* by the Institute for Southern Studies provides ample ammunition for this. The authors ranked all 50 states on 256 indicators of environmental health; Oregon came out at the top, with Alabama at the bottom. (See Table 2-1.) The breakdown of these indicators into environmental policies and conditions reveals interesting patterns across the country. (See Figures 2-1 and 2-2.) As the Institute includes details of which states have done best in all 256 indicators, concerned citizens can point to model programs as they lobby to improve conditions in their own local communities.[12]

Energy and Transportation

We know how to improve the efficiency and sustainability of our use of energy and transportation—that is the bright news when it comes to our future activities in those areas. What remains is for the United States, and other countries, to incorporate that knowledge into policies.

With the 1972 increase in oil prices, market forces rewarded improvements in energy efficiency, and U.S. industry responded. Industrial energy consumption in 1986 was one third lower than it would have been without the 1972 price increase.[13] Energy efficiency is the greatest "source" of new energy available. It cuts the costs that must be carried by energy users and providers, it lowers the pollution that enters the atmosphere, soil, and water, it reduces the extraction of materials from the earth and all the degradation that goes along with it, and yet it maintains society's supply of energy.

In industrial countries, a direct relationship exists between the price of gasoline and the efficiencies of vehicles. With by far the lowest gasoline

TABLE 2-1

STATE	Final Green Index 256 Indicators RANK	Green Conditions 179 Indicators RANK	Green Policies 77 Indicators RANK
Oregon	1	3	2
Maine	2	4	5
Vermont	3	2	12
California	4	19	1
Minnesota	5	5	7
Massachusetts	6	6	9
Rhode Island	7	7	10
New York	8	17	8
Washington	9	13	14
Wisconsin	10	21	6
Connecticut	11	23	4
Hawaii	12	1	24
Maryland	13	14	15
New Jersey	14	28	3
New Hampshire	15	8	20
Colorado	16	10	26
Michigan	17	32	11
Florida	18	30	13
Idaho	19	11	36
Iowa	20	29	16
Montana	21	15	31
Nevada	22	9	43
North Carolina	23	37	18
Delaware	24	27	25
North Dakota	25	16	37
Pennsylvania	26	34	21
South Dakota	27	12	48
New Mexico	28	20	38
Nebraska	29	24	30
Missouri	30	33	23
Illinois	31	42	17
Virginia	32	36	22
Utah	33	22	41
Alaska	34	18	47
Arizona	35	26	39
South Carolina	36	35	32
Ohio	37	46	19
Wyoming	38	25	44
Georgia	39	38	29
Oklahoma	40	31	42
Kentucky	41	39	33
Kansas	42	43	28
Indiana	43	49	27
West Virginia	44	41	45
Tennessee	45	45	40
Texas	46	48	35
Mississippi	47	44	46
Arkansas	48	40	50
Louisiana	49	50	34
Alabama	50	47	49

SOURCE: *1991-1992 Green Index*

FIGURE 2-1: *Environmental Conditions*

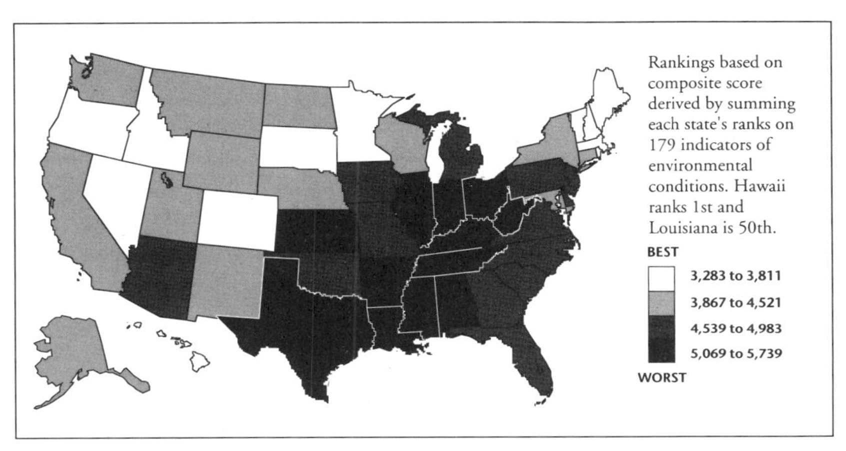

SOURCE: *1991-1992 Green Index*

FIGURE 2-2: *Environmental Policies*

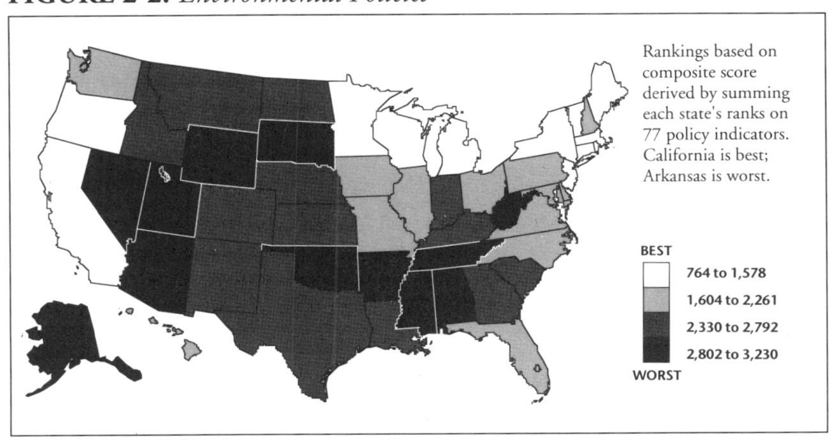

SOURCE: *1991-1992 Green Index*

prices, the United States is also by far the least efficient in miles per gallon. A similar relationship can be seen regarding per capita gasoline consumption. The lowest prices—those in the United States—generate the highest consumption, while the highest prices—in Italy, France, and Japan—generate the lowest consumption.[14] Raising the price of gasoline in the United States is an important first step to improving the country's energy performance.

I'm arguing that there we were with 500,000 troops in the Persian Gulf and all we were able to do was raise the tax on gasoline by a nickel. The Europeans are paying two to four times the amount we do for gasoline. We built our society around cheap gasoline and the car. It has to change.

The trouble is that politicians, unless there is some kind of support for a shift like that, don't want to talk about it. And there isn't much support. It is difficult to talk about a tax that is isolated from anything else. If you talk about a tax shift, maybe you can begin to talk about these things.

We are doing better. There was a debate about a carbon tax in the budget summit last year and I think that that debate will increase. If you want to choose the efficient policy means to reduce carbon emissions, it is the carbon tax—that is the way to go.

Rafe Pomerance, *United States*
World Resources Institute
Earth Summit, p. 180

In New England, the Conservation Law Foundation (CLF) joined with other environmental and consumer organizations as the New England Energy Policy Council, and in 1987 analyzed that area's energy requirements. The council found that through greater efficiency, the region could meet between 37 and 57 percent of its total electricity needs in the next 20 years. CLF has since signed agreements with the utilities that supply two thirds of New England's electricity to participate in the design and implementation of its program. More-efficient industrial motors will be installed, energy-saving equipment will be invested in, and buildings will be retrofitted for efficiency. Homes will receive attention as well as government and commercial structures. CLF says that power supplied through efficiency costs one quarter to one third less than that from new power plants, is less risky, and creates more jobs.[15]

Because states have central functions in regulating utility rates, building codes, appliance standards, mass transit, ride-sharing, management of agricultural and forest lands, energy use in government buildings, and procurement of government equipment, they have major potential to improve energy and transportation policies. California began to review and change its standards for buildings and appliances and its conservation incentives for utilities in the early 1970s. Between 1973 and 1985 it saved $23 billion by reducing by 35 percent the amount of energy businesses used to produce a dollar of goods and services. While the per capita use of energy in the United States was rising, in California it declined. In late 1990, the conservation budgets of state utilities were further increased and least-cost planning was implemented with the goal of meeting almost half the state's new electricity needs in the next 10 years with renewable sources of energy.[16]

Portland, Oregon, developed a joint public-private mass transit plan in the early 1970s to reduce traffic congestion, improve air quality, and encourage its downtown businesses and night life so that the city would not be overcome by growing numbers of shopping malls on its outskirts. It was not intended to deal with problems of climate change or energy, but it has done that as well. A light rail system and buses were set up, and a "parking lid" froze the number of parking spaces. Although the number of people employed downtown has since grown from 69,000 to 90,000, the number of automobile trips has remained about the same, and the number of parking spaces has only been adjusted twice from its original 40,000. Portland is now a model for other American cities struggling to deal with similar issues.[17]

In Congress, a number of bills have affected energy and transportation in recent years. The Global Climate Protection Act of 1987 authorized the President to develop a coordinated national policy on climate change; although the President has not yet taken serious action to do so, such bills can be important in laying the groundwork for change. The Energy Policy and Conservation Act of 1975 mandates many conservation measures ranging from appliance efficiency standards and labelling to Corporate Average Fuel Economy standards. It was weakened, however, by the 1980 Automobile Fuel Efficiency Act and other changes in 1986.[18]

The Public Utility Regulatory Policies Act of 1978 requires electric utility companies to purchase power supplied from certain independent producers, including those that use renewable fuels. The Alternative Motor Fuels Act of 1988 orders feasibility studies of alternative fuels in federal passenger vehicles, as well as assistance to state and local governments for the testing of alternative fuels for buses. This is only an incomplete list, and legislation of all types can touch in minor or in major ways the use of energy and transportation.[19] It can be complex, however, and is always highly political. Many bills fall short of expectations once they are translated into reality.

Along with energy efficiency, another top priority is the development of renewable sources of energy, including solar, wind, and hydropower. Hydro dams and industrial wood burning, along with a small amount of energy from other renewable sources, now meet about 8 percent of total U.S. energy needs. President Bush's National Energy Strategy (NES), released in February 1991, calls for the expansion of hydropower, waste-to-electricity power plants, and liquid fuels from biofuels. But according to James MacKenzie of the World Resources Institute (WRI), it contains no significant initiatives beyond research and development for increased amounts of electricity from photovoltaics, wind, and solar thermal sources. MacKenzie quotes the NES as saying those technologies "have not yet reached a level of performance and cost that permit their diverse, widespread use."[20]

This NES finding is questionable, and much research is available to demonstrate the viability and value of solar power and other renewable sources. Most important, though, the NES statement becomes a self-fulfilling prophesy: if government and industry do not invest in the development of those renewable sources, then they will remain underdeveloped. Commitment is what is most needed in the area of renewable energy. The sun, oceans, and the earth provide the means, and scientists have already made tremendous steps in providing the technologies. What is lacking is policy support. The federal government should follow the leads of states like California, and of many local initiatives, and encourage the types of energy that pollute least, do not draw down energy supplies, and hold the most long-term promise for the future.

To reduce oil use in vehicles, the NES advocates the introduction of alternative fuels, such as methanol, ethanol, compressed natural gas, and electricity. It would extend the ethanol tax subsidy, and it advocates increased research and development for alcohol fuels from biomass and coal. However, it rejects other economic incentives. According to MacKenzie, "economic incentives to encourage efficiency—such as oil-import fees or carbon taxes—are categorically rejected because of alleged potential negative impacts on inflation and economic growth."[21]

The NES sees a complete turnaround for nuclear power, reports WRI. Through regulatory reform, development of new reactor designs, and a vigorous effort to find waste disposal solutions, the NES expects nuclear power to provide twice as much energy in 2030 as it does today.[22] "*Without* successful implementation of the Strategy initiatives, the contribution of nuclear power to our electricity supply could decline substantially after 2010," says the NES. "*With* the Strategy initiatives, nuclear power could be generating, safely and cleanly, as much as 21 percent of our total electricity needs by the year 2030."[23]

The NES projects coal consumption to rise almost 50 percent by 2010 and nearly 70 percent by 2030. Total energy consumption, it says, will continue to rise for at least the next four decades. By 2030 it will have increased by 50 percent above today's levels. By 2010 oil use will be up 12 percent over 1991, natural gas up 22 percent, and coal up 47 percent. Under this scenario, U.S. carbon dioxide emissions will increase 25 percent over the next two decades. (See Figure 2-3.)[24]

The NES has been severely criticized. Objections have targeted the plan's emphasis on increasing the supply of energy instead of using efficiency to reduce demand, its aim of drilling for oil in environmentally sensitive areas, its lack of commitment to funding renewable energy research, its failure to substantially reduce greenhouse gas emissions, and its failure to reflect the true environmental and public health costs of energy, especially nuclear power.[25] The Union of Concerned Scientists has put forward a set of alternative recommendations, with a detailed

FIGURE 2-3: *Trends in U.S. Carbon Dioxide Emissions*

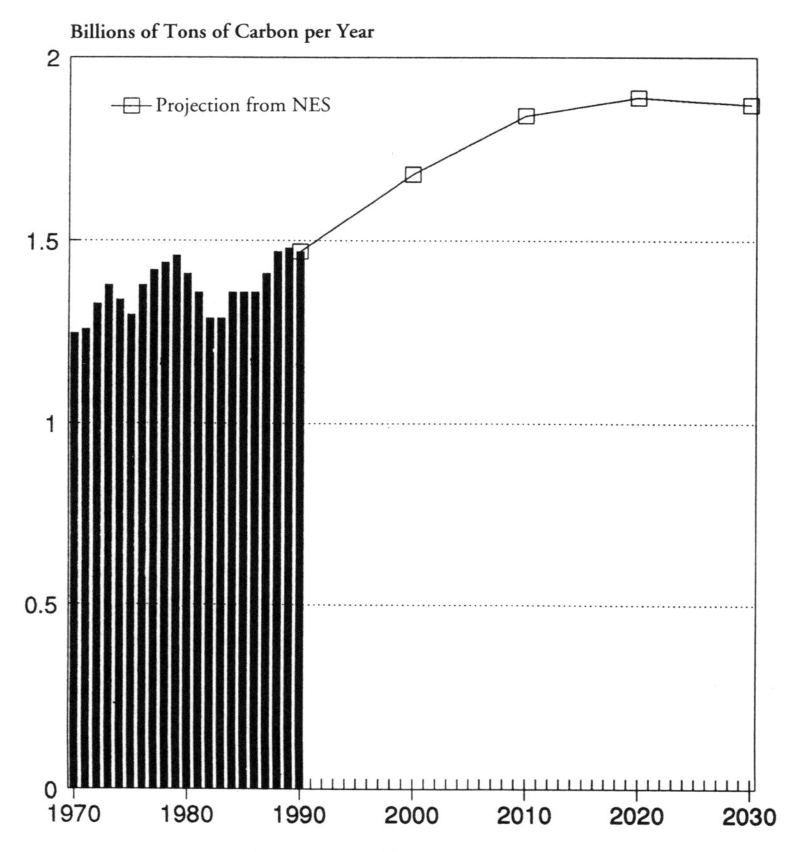

SOURCE: James Mackenzie, World Resources Institute

rebuttal of the NES proposals. Their vision, and that of many other groups working on these issues, is of a country that depends on renewable energy sources, alternative fuels, and energy efficiency—and not on nuclear power.[26]

The existence of such recommendations and the knowledge that enables them hold promise for greater sustainability in future plans. And the creation by the governments of other countries, such as Canada and the Netherlands, of better plans is a signal that awareness of interwoven relationships among energy, environment, transportation, social, economic, and other policies is affecting their planning regarding those subjects.[27]

Technology

Technology is often thought of only in terms of physical devices. It is more useful, however, to view technology more broadly as a treasured ingredient of society's base of knowledge. Attention is often paid to the areas where technology to meet environmental or social goals is lacking, or where industrial processes operate against long-term well-being. But the examples of technological success in day-to-day life and production are many, and often taken for granted. Industry recycles myriad materials; while auto emissions into the atmosphere are nowhere near a sustainable level, they nevertheless have been greatly improved in recent years by technology; and many materials have been developed that reduce the total quantity that must be extracted from the earth.

Along with reduced consumption, new technologies are needed urgently to propel us toward sustainability at a rate faster than other factors move us away from it, and certain areas of research may offer particularly important environmental benefits. One such area is that of "industrial metabolism," a phrase coined by Robert Ayres to address sources of pollution arising from normal use and wear and tear of equipment, such as corrosion of pipes, degradation of painted surfaces, and the wearing down of brake linings. Studies of environmental metabolism move us away from a focus on point-source pollution and toward methods of accounting for all the flows of materials in the environment. For example, while scrubbers can prevent a great deal of sulfur from entering the atmosphere, they simply transfer it to land or water. But "source control," "pollution prevention," and "waste minimization" systems, all of which refer to similar processes, involve process redesign, raw material substitutions, management changes, product substitution, and improved precision and control in order to accomplish environmental goals.[28]

Another important area is for techniques of tracking pollution. The ability to measure its movements into other states, or to measure emissions from the tailpipe of cars on the highway, could have considerable significance for pollution policy. Technologies that reduce demand for energy or water are currently offering impressive savings, but their development has been slowed by such policy impediments as subsidies for water use. (See Chapter 5.) Success is likely soon in the area of improved thermal performance of window coatings, making them superior even to walls in keeping in warmth. Basic research on the physical environment of buildings is greatly needed—energy analysis, indoor air quality analysis, and the ability of buildings to promote human productivity and comfort are all priorities. Much progress in those areas could be associated with the development of new technologies.[29]

Institutional factors, rather than technological impediments, are the greatest barriers to the implementation of cleaner, more-efficient technologies. In a random sampling of 118 clean technology installations by industry in the mid-1980s, only 45 percent relied on new licensed technology, whereas 30 percent were the result of management initiatives using off-the-shelf technology, simple housekeeping improvements and common sense management initiatives.[30]

Old technologies persisting from a time of low resource and waste disposal costs must be replaced with newer, cleaner processes. Market incentives are not strong enough to drive this change except in cases of extraordinary savings, although these certainly can occur. (See Box 2-1.) In most cases, however, waste reduction innovations must compete with other investment possibilities. Both corporate and government policies must be updated to provide economic support to the many statements endorsing the importance of waste reduction efforts.

It is already known which policies are most in need of revision. Policies that require "best available technology" standards can perversely discourage the development of a new technology that might be better. Other policies focus on "end-of-pipe" pollution controls instead of working

Box 2-1: *Clean Technology*
The Eaton Corporation manufactures truck axle housings in its Humboldt, Tennessee, facility. Metalworking processes resulted in the generation of approximately 490,000 gallons of spent fluids yearly that required incineration at an annual cost of $80,000. The equipment was redesigned to permit closed-loop recycling of the process water with the removal of contaminants. Machine oil removed from the process water is sold to a re-refiner for a profit of $9,200 yearly. Reduced fluid make-up results in savings of $102,000 yearly. The cost of the innovation was $158,000, with annual savings of $191,200—a return on investment in less than one year.

SOURCE: Excerpted from Larry Martin, *Proven Profits from Pollution Prevention, Vol. 2,* page 87, (Washington, D.C.: Institute for Local Self-Reliance, 1989).

toward reduction of materials usage overall, while many concentrate on air, water, or land alone instead of working with systems as a whole. And many regulations that require minimum standards do not also contain incentives for going beyond those minimums. New policies are needed that correct these deficiencies, and economic instruments and performance-based standards are needed as well.

Environmental Taxes as Economic Incentives

Tax policy is often used by governments to encourage various types of activities, or to discourage others. Although it is not, it could be used to bring market forces into line with the environmental and social costs of activities that take from the future. Such tax-based price adjustments could improve the functioning of the market by amounting to the inclusion of the true costs of an activity.

Revenue from such taxes could, in some cases, replace that from other taxes that currently give incentives for overuse of the environment. And they could also, temporarily, lower pressures of standard current taxes, which target income, profits, and the value added to goods and services. They are successful in many ways, but they can discourage work, savings, and investment. Environmentally sound taxes might encourage investment in new environmental technologies and open up markets abroad for such new American technologies. And by shifting the tax burden away from income, they might have a stimulating effect on the economy.[31]

"Green taxes" already exist, although most are too small to affect significantly the structure of human activities that are putting stress on the environment. Examples are fees charged on sales of fertilizers and batteries, and various product charges. Norway, for example, taxes fertilizers and pesticides to raise money for sustainable agriculture programs, which is a positive use, but at the moment the charges are too low to reduce chemical use on farms to any great degree.[32]

A higher tax on leaded gasoline in the United Kingdom, however, increased the market share of unleaded gas from 4 percent in April 1989 to 30 percent in March 1990. And recent U.S. legislation to tax chlorofluorocarbons (CFCs) will approximately double current prices, and then double them again by 1995, followed by another significant rise by 1999. The tax is expected to generate $4.3 billion, to encourage production of alternatives to CFCs, and to reduce the use of these chemicals that destroy the ozone layer. (See Chapter 5.) These are among the first significant examples of successful taxes for protection of the environment and public health.[33]

Taxes could be placed on carbon emissions, thereby contributing to commitments at the international level to reduce them. They could be used to penalize the use of virgin materials like the cutting of primary forest, and hence provide incentives for recycling, reuse of materials, and reduction of materials usage. Charges for the generation of toxic wastes could foster waste reduction efforts and the development of safer products. If emissions of chemicals into the air were reduced through a tax incentive, paybacks would come in many forms, including reduced strain on the nation's supply of medical care to people with respiratory illnesses and fewer work hours lost to such diseases. If taxes discouraged the overpumping of water and increased water efficiency, then economic benefits would go to farmers and other people in many areas.[34]

A study by the Congressional Budget Office (CBO) on possible carbon taxes that could be levied on coal, oil, and natural gas found that a major tax phased in between now and the end of the decade would cut carbon dioxide emissions by 37 percent, improve the nation's energy efficiency by 23 percent, and encourage investment in non-carbon energy sources. The

CBO model projects a $45-billion drop in gross national product in 2000, which amounts to a decline of only 0.6 percent, much or all of which could be made up for by a pairing of the carbon tax with reductions in other levies, such as income tax.[35] Based on this and other studies, Worldwatch Institute has produced a sample scenario for environmental taxes.[36] (See Table 2-2.)

"Eco-taxes," as they have also been called, are not without difficulties, however. They would be a regressive form of tax, placing as high a burden on the poor as on the rich, for example, so allowances would need to be made to lessen their impact on the poor. In addition, their revenues often would not be lasting: once industries and individuals switched to new technologies less dependent on carbon or on the targets of other eco-taxes, revenue would decline. Theoretically, some of this decline would be offset by a decrease in the overall environmental and health costs that society has to bear. Nevertheless, eco-taxes cannot replace income taxes, but they could be part of an integrated response through economic incentives and other means to bring activities more in line with natural constraints and make them more consistent with sustainability.

TABLE 2-2: *United States: Potential Green Taxes*

Tax Description	Quantity of Taxed Activity	Assumed Charge[1]	Resulting Annual Revenue[2]
			(billion dollars)
Carbon content of fossil fuels	1.3 billion tons	$100 per ton	130.0
Hazardous wastes generated	266 million tons	$100 per ton	26.6
Paper and paperboard produced from virgin pulp	61.5 million tons	$64 per ton	3.9
Pesticide sales	$7.38 billion	half of total sales	3.7
Sulfur dioxide emissions[3]	21 million tons	$150 per ton	3.2
Nitrogen oxides	20 million	$100 per ton	2.0
Chlorofluoro-carbon sales[4]	225 millions kilograms	$5.83 per kilogram	1.3
Groundwater depletion	20.4 million acre-feet	$50 per acre-foot	1.0

[1]Charges shown here are for illustration only, and are based simply on what seems reasonable given existing costs and prices. In some cases several taxes would exist in a given category to reflect differing degrees of harm; the hazardous waste tax shown, for instance would be the average share. [2]Since revenue would diminish as the tax shifted production and consumption patterns, and since some taxes have multiple effects, the revenue column cannot be added to get a total revenue estimate. [3]The Clean Air Act passed in October 1990 requires utility sulfur dioxide emissions to drop by 9 million tons and nitrogen oxide emissions by 1.8 million tons by the end of the decade. [4]This tax already exists. Revenues shown here are expected for 1994.

SOURCE: Worldwatch Institute.

CHAPTER 3

The International Arena

The world stage of international law, institutions, and government relations holds impressive potential to pull countries forward toward the sustainable use of world resources and investments in people and their future. It often seems slow to change and full of inertia, but the international arena is full of possibilities and constructive language, and is one of the most innovative forums for confronting the problems that face us. For decades the nation-state has been the focus of efforts to manage human affairs. Now it is becoming clear that international and regional bodies, along with the groups working for local self-reliance discussed in Chapter 2, hold unrealized opportunities for effective change.

The international arena is a framework in which the successes of any one country can spread to the others, the needs of one can be met by the products or skills of another, and the efforts of many can join together in a unified direction. Its structure is built of international organizations and of relations between countries, with basic patterns of trade, population growth, human development, and environmental change underlying it.

Human Rights

At the foundation of progress toward sustainable development are human rights. They include the right to political participation, economic participation, and freedom of speech; the right to work and to rest; the right to education; and the right to participate in the cultural life of a community.

These are all part of the United Nations Universal Declaration of Human Rights.

Countries where people lack these rights are those with the most severe environmental degradation, and in many cases also those with the most severe human deprivation through poverty, war, and government terror. Countries where people enjoy these rights are those now most responsive to the public desire for a clean environment and for political liberties.

They rest on another series of rights also contained in the U.N. Declaration: the prohibition of torture; equal protection of the law; an effective remedy for violations of rights; prohibition of arbitrary arrest, detention, or exile; the right to a public hearing by an independent tribunal; freedom of peaceful assembly and association; and limits on governments.

Global improvements in human rights have been steady since World War II. Before then, international mechanisms for their promotion did not exist, and they were neither an international goal nor a diplomatic concern among nations. But today, international participation and aid are often tied to improvements in human rights, and the world community is often able to affect the rights of people inside national borders despite their formal inability to specify the way a government treats its people.

The United States has often been a leader in the area of human rights, as have such non-governmental organizations (NGOs) as Amnesty International. They have defined human rights as a precondition for positive national development, and correctly so, although in some cases governments may have done so for ulterior motives. Positive language often has come from U.S. diplomats regarding human rights in international forums.

At a major international meeting on the protection of the environment in Sofia, Bulgaria, in 1989, the head of the U.S. delegation, Richard J. Smith, said:

> Let me emphasize that we [the U.S. government] see a key linkage between environmental progress and freedom of expression. Individuals and organizations must be free to express their environmental concerns and press to have them addressed or environmental problems will not be solved. This is a central message that we need to bring

> to Sofia....We would have reservations...about any final document that did not include provisions on human rights questions, including particularly aspects of the environment (such as rights of individuals to information on environmental matters and strengthening the rights of environmental activists).[1]

These statements came in the wake of violations of the human rights of Turkish and Muslim minorities in Bulgaria, and of efforts by NGOs to gain entrance to the Sofia meeting that were repelled by the Bulgarian government. The NGOs were later instrumental in the organization of the groups that peacefully brought down the Bulgarian government, as well as those of some of other East European countries. The U.S. government had long supported the democracy movements of the East European people as a means of fighting the Warsaw Pact governments, and human rights and environmental concerns became the means to that end.

Yet the government was also sincere. Long-lasting peace, democracy, and cooperation in that region and others are tied to human rights and the desire and need for a clean, productive environment. Smith said in addition "we must emphasize that the major environmental problems are not national, but international in character. Pollution knows no national boundaries; insisting that these issues be addressed cooperatively does not constitute interference in internal affairs. Rather, such insistence is the only way real progress can be made."[2]

For many people, the connection between human rights and environmental goals is increasingly clear. (See Box 3-1.) And the link was pointed out 20 years ago in the Declaration of the United Nations Conference on the Human Environment in Stockholm: "Man has the fundamental right to freedom, equality and adequate conditions of life, in an environment of a quality that permits a life of dignity and well-being, and he bears a solemn responsibility to protect and improve the environment for present and future generations. In this respect, policies promoting or perpetuating apartheid, racial segregation, discrimination, colonial and other forms of oppression and foreign domination stand condemned and must be eliminated. The natural resources of the earth...must be safeguarded for the benefit of present and future generations."[3]

Box 3-1: *Human Rights and the Environment*
On June 5, 1989 Fang Lizhi, an astrophysicist and a life-long outspoken advocate for democratic reform in China, sought and received refuge in the U.S. embassy in Beijing, where he and his wife lived for a year until the U.S. Ambassador was able to negotiate their departure from China. In the preface to a collection of Fang's writings on China he says:

Right now humanity increasingly faces problems of a global nature: population, energy, environment, atmospheric warming, deforestation, and so on. But as long as there are governments in the world that can hold up the slaughter in Tiananmen Square as a glorious achievement, as long as there are dictators who refuse to be constrained by universal standards, it is hard to imagine that there could be the necessary understanding and cooperation to solve global problems. On the contrary, there have long been precedents demonstrating that the appeasement of governments which revel in slaughter is an invitation to worldwide catastrophe. Because of this, human rights are a global problem, maybe even the most important one. Without steady progress in the human rights environment all over the world, it will be very difficult to find serious solutions for other environmental problems in the global village.

In this sense, China's problems are *global problems. The earth is a small place and getting smaller. What destroys the environment in one part affects the rest. In the same way that the meltdown at Chernobyl contaminated the atmosphere of half the planet, the massacre at Tiananmen Square contaminated the human rights environment of the whole planet. Anyone with global concerns cannot fail to be concerned about this.*

SOURCE: Excerpted from Michael J. Kane, *International Protection of Human Rights and the Environment,* U.S. Environmental Protection Agency, May 1991, unpublished paper.

The United Nations Conference on Environment and Development, on the twentieth anniversary of the Stockholm Conference, brings the U.N. commitment to the connections between human rights and environment up to date. It is hoped that the heads of state and government in Rio will endorse an Earth Charter on environmental rights and responsibilities to parallel the U.N. Declaration on Human Rights. Alternatively, perhaps they should just consider the right to clean air, water, and so on a basic human right, and so amend the earlier declaration.

The connection between human and environmental rights is being made at a time when totalitarian governments have fallen and the prospects are brighter than ever for human rights worldwide. With the demise during the late 1980s and early 1990s of oppressive governments around the globe, a focus on human rights is a healthy movement on the way to sustainable development in the international system.

Industrial Country Forums

At a meeting in Helsinki, Finland, in 1975, a floating set of negotiations in which 35 countries try to work out pan-European rules for human rights, military behavior, economic systems, environmental protection, and other topics was created. The Sofia meeting mentioned earlier was part of that process. So were the Helsinki Accords, one of the earliest and probably the most successful international agreements on human rights.

The Conference on Security and Cooperation in Europe (CSCE), as it is called, probably deserves more credit than it gets for progress that has occurred in the areas just discussed. It is a mechanism for consultation that has sometimes succeeded where more rigid and structured forums have not. With no permanent offices and no significantly fixed institutional shape, it is easily adaptable to fit whatever circumstances are needed for progress.

Especially in the last two years, the United States has joined the Soviet Union and Germany in putting a high priority on the CSCE. When high-level summit discussions in 1989 did not reconcile differences among countries, the United States decided to put much more effort into lower-level CSCE talks. The *Wall Street Journal* reported that the United States wanted the CSCE negotiations to lock in such reforms as free elections and economic liberty in Eastern Europe and the Soviet Union, as well as to expand the Helsinki human rights declarations to define free, multiparty elections as a basic human right.[4]

German Foreign Minister Hans-Dietrich Genscher has argued that NATO will decline in importance in coming years and the CSCE will increase.[5] It bridges the gap between NATO and the former Warsaw Pact countries, as it includes them both. And it allows environmental, social, military, and other concerns to be treated by a single process rather than dividing them among NATO, the United Nations, the European Community, and other organizations.

The Ford and Carter administrations were attacked for adhering to the Helsinki process by critics who argued that it did little to guarantee human rights, and that it implied U.S. recognition of Soviet control in Eastern Europe. But they have been proved wrong, as the Helsinki rules have been cited repeatedly in the East as a reason for reforms there.[6] With the end of East-West hostilities, the CSCE offers major potential for international cooperation on exactly the issues required for sustainable development. It is a healthy adaptation to changes in the international environment that raised the need for new institutional forums.

Another institution with potential to confront sustainable development issues is the Organisation for Economic Co-operation and Development (OECD). Focussing mostly on economic issues, it brings together representatives of the 24 richest countries to search for agreement on problems. Although it regrettably lacks significant input from the rest of the world and non-governmental organizations, the OECD may be the most important forum for the countries that are the largest producers of greenhouse gases, hazardous wastes, and pollution, and that are the biggest consumers of energy. If the OECD endorses changes in national behaviors regarding the environment, that would influence national actions.

Many of the most fundamental economic issues are debated at the OECD. It has recently taken up, for example, the topic of trade and environment; its two directorates on these topics produced a joint report that says "trade and environmental policies should be seen as being mutually supportive rather than in terms of conflicting interests....Unlike sustainable development, free trade is not an end in itself; it is rather an important means for achieving economic efficiency and growth." Coming from a group of economists, that is a significant concession, and one that makes economic goals subservient to the larger goals of sustainability and development.[7]

If the OECD reaches a consensus on trade and environment topics, then its member countries will likely incorporate its findings into the General Agreement on Tariffs and Trade (GATT) in the years ahead. GATT, the world trade regime, is by far the most important agreement regarding commerce. It is generally not open to advice from the public, and can be an extremely difficult organization to influence. The OECD is an important international forum whose decisions on trade and environment could find their way into future GATT negotiations. As such, the start of discussions on this topic within OECD is important.

The OECD also considers such topics as environmental accounting, and if it were to recommend changes in national income accounting systems, its members would listen. Even though the OECD cannot compel any country to follow its advice, as an institution it has impressive potential to sway the beliefs of otherwise resolved nations and is highly respected when it speaks out on economic or other topics. Increasing the emphasis it puts on investments in environmental, social, and sustainability goals would give a major boost to balanced international development.

Of course, all international institutions are subservient to the nations within them, and they will continue to be limited by the agendas of those nations. Nevertheless, they can serve as a forum for change and improved sustainability in development.

The United Nations

The United Nations, unlike the CSCE or the OECD, has almost every country on earth as a member. It has a large number of organizations that promote sustainable development in many different ways. But it also has a great deal of bureaucratic inertia and inefficiency. The politics of the member countries play themselves out in the organization as well, and that can distort decisions away from sustainability. Overall, it is both a huge asset to sustainable development and a barrier when it finds itself unable to climb the fences that international politics erects.

Much work is going on today to find ways to improve the performance of the United Nations, especially given its important role in the Persian Gulf war. The U.N. Conference on Environment and Development is thus occurring at an opportune time for reform within the U.N. The goal must be to remove barriers the organization has long faced without introducing other institutional problems in their place.

Some proposals simply call for strengthening existing parts of the

United Nations, most notably the U.N. Environment Programme (UNEP). UNEP's budget was tiny from its creation until 1990, when it received a considerable increase in funding. Even with the additional funds, however, it still has a smaller budget than many U.S. environmental groups. If the United Nations community is to get serious about sustainable development, UNEP must receive significantly more money as well as increased staff and authority.

Other proposals include the creation of new branches of the U.N. One is for a World Environment Authority to administer new conventions dealing with global warming and climate change. Another is for an Environmental Security Council that might mirror the existing Security Council but work on environmental goals. Alternatively, the current Security Council could be transformed to include environmental concerns. Creating an International Environmental Tribunal is another possibility; it could offer rulings over environmentally related topics and maybe create a list of precedents to govern future behavior. An Environmental Ombudsman could be created, an Environmental Chamber could be added to the International Court of Justice, and an Environment Commission or a Sustainable Development Commission similar to the Human Rights Commission or the International Labour Organisation could be created.

And there are many proposals to adapt existing U.N. organizations to make them better fit environmental needs. The Economic and Social Council could be reformed to include a high-level Coordinator for Environment and Development, who would have oversight and authority over all U.N. agencies. The moribund Environment Coordination Board under the Chairmanship of the Executive Director of UNEP could be revitalized. UNESCO could be given the mandate of assisting member states in building scientific capabilities for environment and sustainable development efforts. Or the Trusteeship Council of the U.N. could be transformed into an environmental chamber. Whatever happens, a reformed or an entirely new U.N. body could face the challenge of devising and administering such innovations as a proposed credit program for carbon emissions, in which countries with high emissions purchase from developing countries credit to reduce their output more slowly.

Strengthening the international legal order is the focus of the Declaration of the Hague on the Environment, the Draft Economic Commission of Europe, Charter on Environmental Rights and Obligations, and the recent Stockholm Initiative on Global Security and Governance.

Related changes could take place within nongovernmental organizations that would parallel the U.N. changes. An Amnesty for the Earth organization could be built, modelled after Amnesty International, to investigate environmental violations and issue an annual report on the environmental records of governments. Independent scientific bodies, such as the International Council of Scientific Unions, could be involved in states' efforts to comply with international conventions. In all cases, NGOs are bound to be more involved, formally and informally, in any U.N. efforts in these fields, building on their unprecedented inclusion in preparations for the 1992 U.N. conference.

At the governmental level, the proposals for U.N. reform raise vexing questions about national sovereignty and autonomy. Some people are concerned that trying to let the U.N. carry out environmental monitoring will actually reduce support for the organization as a whole by those members who feel such a move impinges on their national sovereignty. Others call for even stronger institutional authority to address urgent environmental problems. Also unresolved are questions about priorities: How do member states expect U.N. agencies to balance the right to a clean environment with the rights to use local natural resources for food, shelter, energy, and so on?

Whatever the outcome of the Rio conference, the meeting is only the starting point for the continuing evolution of legal instruments and global institutions to address the many pressing issues summarized in the term sustainable development. On the whole, possibilities for stronger abilities on the part of the U.N. to meet environmental needs are many and highly plausible. Similar initiatives will be needed to address such other needs as international public health, education opportunities, and personal security. Those goals probably have as much potential as environmental protection does for strengthening nations and people worldwide when taken up more actively and more effectively by international institutions.

All such initiatives will depend on active participation in their development, especially by the American people and government, and the President and the Cabinet should be involved directly. In light of growing and changing stresses, the healthy response will be to invest heavily in new initiatives to deal with new problems.

Specifically, the United States should work toward new treaties and improved and possibly new institutions. The United Nations Association and the Sierra Club have listed some of these initiatives. They include early negotiations to establish a global warming pact, plus a two-step process for controlling greenhouse emissions through a cap on and then rollbacks of current emission levels and agreement on a second phase of reductions to fit a global "emissions budget" in the early twenty-first century. Their report, entitled *Uniting Nations for the Earth,* also contains detailed recommendations on strengthening the institutional machinery that underpins all these efforts in the international arena.[8]

The Corporate World

A few corporations have been among the institutions trying to demonstrate healthy responses to the stresses that face international systems. Through the International Chamber of Commerce (ICC), a number have joined in establishing a Business Charter for Sustainable Development. "The objective," notes the ICC, "is that the widest range of enterprises commit themselves to improving their environmental performance in accordance with these principles, to having in place management practices to effect such improvement, to measuring their progress, and to reporting this progress as appropriate internally and externally."[9]

Among the principles are "to recognize environmental management as among the highest corporate priorities and as a key determinant to sustainable developments...to modify the manufacture, marketing or use of products or services...to prevent serious or irreversible environmental degradations...and to promote the adoption of these principles by contractors acting on behalf of the enterprise, encouraging and, where appropriate, requiring improvements in their practices."[10]

The response of the corporate world to the new reality of how to conduct business is in good part often a reaction to goading by NGOs and pressure groups around the world. In the United States, for example, the Coalition for Environmentally Responsible Economies published the Valdez Principles in late 1989 and challenged companies to be evaluated by its 10 broad standards on the effect of corporate activity on the biosphere. Although by mid-1991 only 21 companies had signed up and agreed to adhere to them, the Valdez Principles have been an effective tool for raising stockholder and corporate awareness of the issues.[11]

Many of the actions needed to protect the environment are the same ones that corporations have traditionally depended on for increased profits: improving efficiency, developing processes that use fewer materials, and reducing waste. In that sense, environmental goals are consistent with the profit motive, and strict environmental regulations can lead to healthy innovation. Tapping this potential is now often as much the responsibility of a corporate Vice President of the Environment as it is of the Product Development Manager.[12] To strengthen this approach further, each line manager from the chief executive on down should think of himself or herself as an environmental manager. In addition, clear accountability for environmental performance should accompany managerial responsibility.

Businesses also benefit from environmental improvements when consumers choose "green products" over ordinary ones, something all polls indicate they want to do. Ninety percent of Americans say they would pay more for green products, and nearly half of them claim to have taken steps to be more "green," such as avoiding aerosol sprays, reducing their use of paper towels, or seeking out products made from recycled materials.[13] A survey of 80 major companies found that more than half believed greater consumer concern about the environment had led them to make corporate changes. The accounting firm of Deloitte & Touche, which did the research, concluded that "consumer pressure, regulatory compliance, and the necessity of maintaining a healthy corporate image are changing the way America does business."[14]

Aiding Development

Unfortunately, development assistance from industrial to developing countries has often served various political and business goals of the donor first, and the needs of those who should be its targets—the poorest of the poor—last. As much as two thirds of the aid some governments offer is tied to the purchase of goods and services from their own countries, which is essentially a form of export promotion.[15] The futility of the traditional approach to overseas aid is finally becoming clear as more and more people sink into poverty. (See Box 3-2.)

Box 3-2: *Aid's New Agenda*

What is really needed in the 1990s is a new human order. Its starting point would be the people in each country and the aim would be to improve their conditions, especially the plight of the poor people. Mobilizing international support on this basis could guarantee that international cooperation does not transfer resources from the poor in the rich countries to the rich in the poor countries.

This new human order would recognize that we are all one community on an increasingly crowded planet. This interdependence implies more than economic links. Environmental threats respect no national boundaries—and poverty is the driving force behind many of them. Nor can the international peace process be limited to that between East and West—violence can erupt in any part of the world and affect us all....

In developing a new human order, all partners must recognize their obligations. Developing countries must recognize that much of the responsibility rests on their own shoulders. They can expect international support, but there is no alternative to a sensible restructuring of their priorities.

The industrial nations must recognize that they, too, have urgent problems of poverty at home. But they must also see that improving the human condition all over the globe is in their own interest.

SOURCE: Excerpted from U.N. Development Programme, *Human Development Report 1991* (New York: Oxford University Press, 1991), pp. 78-79.

Development assistance often can achieve its objectives most effectively by investing in relatively small projects to reach local people, and by including appropriate technology. Examples of projects of this type can be taken as prototypes for future projects that will relieve many of the stresses that currently work against sustainable development.

One such project involved the planting of windbreaks in an area that had been deforested by a relentless need for fuelwood, fodder, and construction materials. The Majjia Valley of Niger, the site of the project that was carried out by CARE, was losing almost 20 tons of topsoil per hectare every year from its fields to wind erosion, compared with typical losses of 0.5-2.0 tons per hectare a year.[16]

The trees planted on over 3,000 hectares by the CARE project cut the wind velocity near the ground by 45-80 percent, resulting in less soil erosion and more soil moisture. One study of crop yields found a net increase of 15 percent, and another found jumps of 18-23 percent. The local farmers there are now receiving the delayed benefits of the increased wood production, and can also cut some of the planted wood to use for fuel or construction.[17]

That project illustrates some of the success that activities having to do with the environment can have in improving the incomes and well-being of local people. It is also one of many instances of an NGO having success with the type of local, community-related projects that suit them best. Large development institutions, like the World Bank, have begun during the past few years to recognize the potential of NGOs and to funnel a bit more development money through them. CARE is joined by tens of thousands of community-based and international NGOs that are able to help in all kinds of projects.

By the late 1970s, corn yields in a part of Honduras had declined severely, mostly from the loss of topsoil. People were forced to flee the region and search for livelihoods elsewhere. In 1981 a technical assistance program was started by a private Honduran group, the Association for the Coordination of Development Resources, and World Neighbors, a development group based in Oklahoma. It was oriented toward simple technologies, such as contour and drainage ditches, contour grass barriers, and rock walls, and it included education in fertilization methods involving chicken manure, the intercropping of plants, and limited chemical fertilizers.[18]

The yields of farmers adopting the techniques tripled or quadrupled in the first year, and within five years 40 villages had asked for similar training. All costs of the programs were carried by the farmers themselves, so no donations or subsidies were required, and the payoffs now cover much of Honduras. Unlike the programs of multilateral development banks, the educational techniques used by World Neighbors were carried out in the fields with hands-on activities by the farmers themselves. Those who had learned the program taught it to their neighbors. Migration away from the region has largely been reversed. Soil has been reclaimed to the benefit of formerly landless farmers. The agriculture has proved to be sustainable, and its balanced nature reduces pressures on forests and rivers.[19]

A project by Appropriate Technology International (ATI), a development assistance group in Washington, D.C., has sought to provide incentives for reforestation and more careful use of forests and wood in Guatemala. With ATI support, a local wood workers' association is improving its technologies for furniture making and other items by bringing in new power tools and new drying technology. By increasing the economic value of the wood, Guatemalans have been given a reason to treat it as valuable.[20]

Such projects often illustrate that development is something that is carried out best by the people who are intended to benefit from it, not by bureaucratic institutions in wealthy countries. When local people assist in the planning, projects tend to be more successful than when the affected people are excluded, as is most commonly the case. Development assistance must suit the cultures, goals, and resources of the people in the regions where the development is to take place, not of people who live far away. And no economics or other textbook or course can teach what those are. The encouragement of more projects of the type that have already succeeded will be a healthy response to the stresses that have so far blocked international development.

Underlying Systems: Population and Trade

A healthy response to stress in the international arena must include working to head off pressures right where they begin—with rapid population growth and with international economic relations that trap many countries in downward cycles of poverty.

In 1984 the United States reversed its long-standing position as a leader in efforts to improve worldwide access to family planning services. At the World Conference on Population in Mexico City, the government announced its new position of not funding any organization that counsels women about abortion, and withdrew its funding from the world's two most effective family planning organizations—the International Planned Parenthood Federation and the United Nations Population Fund. Representative Chet Atkins of Massachusetts recently called this decision a "perverse combination of twisted demographic logic and misdirected morality."[21] Although the United States is still the largest provider of contraceptives in the world, through the Agency for International Development, the Mexico City policy may be the biggest barrier to population stability worldwide.

As long as population growth outpaces governments' ability to extend basic immunizations, clean water, sanitation, education, and clean housing, population growth will limit human opportunities rather than add to them. Children should be our greatest asset, not a stress on society. But for them to be so, couples must have the ability to plan their families. Currently, more than 300 million couples say they want to space or limit the number of children they have but lack the means to do so.[22] Women's education and couples' access to a full range of family planning services and birth control are crucial to changing this.

Bright examples of success in family planning do exist. A family planning program in Zimbabwe, for example, has surmounted cultural and economic obstacles to provide services to women. It now has a network of 637 community-based distributors of contraceptives and advice who travel door to door by bicycle and provide materials free to people with low incomes. Most of the women accepting the services began contraceptive use to increase the spacing between births, which has important effects on the health of mothers and on fertility rates.[23]

Even though this program may seem local and not of relevance to Americans, it is directed at the source of many global stresses and represents a healthy response to them. As the most powerful government on earth, the United States should follow its lead. By turning families and human resources into assets instead of liabilities, many other stresses can be attacked as well. This is an underlying theme of many organizations recommending changes for the future.

For example, increasing U.S. support for family planning is one of 12 recommendations made in 1991 by a high-level Task Force on International Development and Environmental Security set up by the Environmental and Energy Study Institute of Washington, D.C. Another is to make the General Agreement on Tariffs and Trade more responsive to environmental needs and objectives.

GATT was written at a time when environmental considerations carried less political weight and were not seen with the same urgency they are today. It contains no provisions for the environment in its articles, even though trade is inherently about the movement of goods that come from the natural resource base. And an August 1991 ruling by a GATT deliberative panel that the U.S. Marine Mammal Protection Act cannot limit U.S. imports of Mexican tuna for conservation reasons suggests that current GATT articles cannot be used to support environmental needs.[24]

Healthy responses to the problem had been proposed even before the tuna ruling—the next round of GATT negotiations may be an environmental one, and may add environmental or sustainable development clauses to the agreement. However, current environmental discussions in the GATT are focused on how environmental measures could distort trade. Global environmental/trade policy must, to the contrary, consider how trade distorts the environment. The GATT may not be the appropriate international body to make this determination. Environmental regulation of trade should certainly be a priority in Agenda 21 and any environmental treaties.

Healthy responses will face impressive difficulties. Much resistance will be found to the introduction of environmental goals into trade agreements, from rich countries and even more so from poor ones. Environment and equality must go hand in hand at GATT negotiations. And environmental goals will likely only be effective if they are matched by other changes that enable the poorest countries to meet such imperatives without neglecting people's day-to-day needs.

Both goals might be met by increases in commodity prices for the products of poor countries, most of which are based on natural resources. If timber, for example, were priced for trade based on the value of all the services a forest can provide for the future, the present, the ecosystem, and society, then its price would rise. This would generate greatly increased income for the country that exported it and allow that country to sell less timber while bringing in the same amount of income, thereby reducing overexploitation of the environment. It seems like a case of solving two overwhelmingly pressing problems with one international agreement.

Numerous other issues within the GATT also have major environmental impacts: quantitative restrictions, dumping, intellectual property rights, and harmonizations are just four of many major trade policies relevant to the environment. These would be unprecedented changes, and they touch on areas and politics that have encountered huge resistance in the past. The measure of how healthy a response the international system will be able to make to environmental and social stresses may be how well countries can adapt the GATT to these new imperatives.

To help nations develop, income from trade is more effective than income from aid. It carries no interest, does not have to be paid back, and comes with fewer strings attached. It also creates jobs and lets people practice and use their skills, and is more beneficial to self-respect and self-reliance. Rates of population growth decline when people have more control over their lives, more income, and more skills. It is in the poorest and underdeveloped areas that population growth is highest. Hence, higher commodity prices and changes in regulatory policy that help countries benefit from trade may be among the strongest actions for achieving development and reducing rapid population growth. As it is also consistent with environmental protection, it is truly a healthy response to stress.

PART II

Systems Under Stress

CHAPTER 4

Economic Systems

As populations grow and industry increases, the stresses that affect society and the environment become more powerful. When Adam Smith, John Maynard Keynes, and other economists formed that discipline, they may not have imagined the power that the systems they were describing would come to have. Many of their assumptions rested on a view of human capabilities as smaller and of nations as more isolated than they are now.

But today, international trade and finance cut through national borders in a way that no military can, and people's livelihoods and security change due to flows of financial assets and goods. Yet human well-being is tied to these forces in complex ways: no simple relationship exists to explain their connections, and different people are affected in different ways.

Economic growth is often treated as a goal itself, rather than as a means to an end. But the end must be defined in terms of human beings, and their current and future quality of life and security. In a changing world, tools like economic growth and military spending no longer accomplish the same ends they once did. And other tools become more useful than they have been in the past for improving the standard of living and level of security.

The economic growth and the allocation of economic resources that allowed the development of modern hospitals and sanitation may now, instead, add to the human health problems that those hospitals must treat.

And the accounting systems and budgeting processes that have been used successfully for decades are inadequate to the task of providing the insights and information necessary for designing new priorities of sustainable development, including environmental protection and programs to deal with poverty.

Economic Growth Is Not the Goal

Economist Herman Daly has written that "further growth beyond the present scale is overwhelmingly likely to increase costs more rapidly than it increases benefits, thus ushering in a new era of 'uneconomic growth' that impoverishes rather than enriches."[1]

That is a major statement, and one that would require a new conceptualization of what economics is about as we approach the twenty-first century. The idea that people would benefit from limiting economic growth has not been well accepted in the past, nor even significantly debated. The idea that growth can impoverish people will require a redefinition of economics and its goals.

The growing degradation of the environment, and the increasingly regional and global nature of that degradation, offers a graphic view of the reality of environmental limits in the absorption of human activities, particularly if those activities make few adjustments for environmental constraints. The expanding and decrepit squatters settlements in the developing world offer visual proof of the limited ability of communities and cultures to survive in the face of economic power and change. And the consumption-based society of the United States, where people sometimes define their self-worth in terms of possessions and consumption, and from which garbage barges like the one from Islip, New York, sail forth in search of new places to dump their cargo, exemplifies the danger of those limits.

If the western definition of economic growth is at odds with the constraints of natural systems, then once it crosses a certain threshold it will be in conflict with those systems. People's quality of life, and the security of their livelihoods, will no longer increase steadily, but may instead decline, as Herman Daly pointed out. (See also Chapter 2 for a discussion of how that is happening in the United States.)

In many ways, that threshold has already been crossed. What we took for granted a few decades ago, we no longer have: clean air to breathe in the cities, chemical-free food, a community structure where we know our neighbors, and a definition of self-worth based on ourselves rather than what we own. But the most dangerous collapse in well-being would come from such disasters as global warming, ozone depletion, an inability to provide enough food for the global population, and epidemics of disease in overcrowded and desolate urban areas. Indeed, in the poorest countries, malnutrition and epidemics are already facts of life.

Economists are fond of saying that "a rising tide raises all boats," to show that global economic growth will ultimately contain benefits for all people in all countries. But this phrase ignores the fact of consumption of the natural capital on which all life depends for basic sustenance. The tide can rise to a certain level, but then it can recede.[2]

An alternative concept of what the tide represents must be developed. The nature of the tide must be considered: Is it necessary that it rise in the sense of getting bigger, or could it rise through qualitative improvements that increase well-being yet do not increase size? That is the sort of question economists must ask. Not how to make more, but rather how to better meet human needs and sustain the environment and other systems on which people depend.

Economists have often measured success by quantity. After all, it is difficult to measure success in terms of attributes that cannot be measured: How can well-being by counted? What numerical value belongs on a person's health, or his or her ability to practice certain skills, or the satisfaction found in looking at an old-growth forest? Yet promotion of these must be among the goals of economics. If economics measured them rather than simple quantity of goods, it would have less tendency to push against the limits of growth. The notion of limits requires a new set of goals for the discipline of economics. (See Box 4-1.)

At its most basic, economics could be defined as the way that people use the environment around them to meet their needs and desires. Yet traditional economic theory has often defined the environment and even much of what describes people—culture, public health, education, population size, and more—as "externalities." It is a denial, of sorts, of what economics itself is. Economics is about people and their environment, but it discounts those elements in its calculations and formulas.

Proper Resource Accounting

The beginnings of an adjustment in economic priorities may come with a

Box 4-1: *New Economic Goals*
The GNP becomes an obsolete measure of progress in a society striving to meet people's needs as efficiently as possible and with the least damage to the environment. What counts is not growth in output, but the quality of services rendered. Bicycles and light rail, for instance, are less resource-intensive forms of transportation than automobiles are, and contribute less to GNP. Yet a shift to mass transit and cycling for most passenger trips would enhance urban life by eliminating traffic jams, reducing smog, and making cities safer for pedestrians.

Likewise, investing in water-efficient appliances and irrigation systems instead of building more dams and diversion canals would meet water needs with less harm to the environment. Since massive water projects consume more resources than efficiency investments do, GNP would tend to decline. But quality of life would improve. It becomes clear that striving to boost GNP is often inappropriate and counterproductive. As ecologist and philosopher Garrett Hardin puts it, "For a statesman to try to maximize the GNP is about as sensible as for a composer of music to try to maximize the number of notes in a symphony."

SOURCE: Excerpted from Lester R. Brown et al., *Saving the Planet* (New York: W.W. Norton & Co., 1991), p. 124.

change in the way that the United States and other countries measure their national assets.

Each country keeps a tally of its production, consumption, and gains or losses of capital—the gross national product (GNP). Decisions are made based on those measurements. But exactly what do they measure? What do they include and what do they exclude? The GNP measurements used by the United States and all other countries today include the depreciation of machine capital, but not that of environmental capital, or declines in levels of education or health.

For example, the cutting of a forest may exchange a renewable resource, which could offer payoffs far into the future, for a one-time gain. In the GNP, the one-time benefit would show up, but the long-term loss would not. Nor would any aesthetic losses, damage to the larger ecosystem, and diminished roles in regional weather and other patterns.

As a result, decision makers have less means and certainly no encouragement to promote the maintenance and generation of the environmental capital that plays a central role in supporting industry, people, and economic growth itself. Likewise, the importance of increasing levels of education and public health in generating "human capital" are discounted, even though they are as fundamental as "machine capital."

Anything that cannot be represented by a numerical value is excluded from GNP. Moreover, future values and future productive capacity are generally left out as well. And the interconnected roles that resources play in supporting other assets and activities also often do not find their ways into the calculations. Improvements in the quality of products produced does not cause an increase in the calculation of GNP. Improvements in the safety of products, and reductions in the pollution or waste they produce, do not show up in GNP figures. And changing priorities—say, to invest in educating people instead of in mechanical productive capability—may enhance (or reduce) long-term economic and human outlook, and yet do not show up clearly in GNP.

The production of "goods" increases GNP, as well it should. But so does the production of "bads." For example, the sale of a car adds to GNP, but so does a highway accident involving that car—by adding the amount of the hospital bills of its occupants and its repair bills, even though great harm has occurred. The work of a police officer in protecting society is valued in GNP by the amount of his or her salary, and indeed the service to society may be much greater than that amount. And the actions of an oppressive police force operating through fear, like those of many totalitarian governments, also contributes to GNP at the rate of its salaries, although its value to society may be negative.

Among all the components of GNP, that contributed by waste disposal exemplifies misguided resource accounting. In 1986 the congressional Office of Technology Assessment estimated that national spending for waste management and pollution control was about $70 billion, approximately 1.9 percent of GNP.[3] Although environmental advocates tout this contribution to the GNP as important to the economy, in effect it represents a cost to business and the consumer. According to the American Council for Capital Formation, environmental regulations in the 1960s and 1970s led to a 2.6-percent reduction in the GNP.[4]

With proper market incentives, business would shift the allocation of investments from virgin resource exploitation to more durable products, less disposability, and resource recycling. Instead of spending $5-6 billion to handle our nation's municipal solid waste, the funds could instead be channeled into constructive enterprises.[5] It would still be counted in the GNP, but it would be serving both social and environmental objectives, rather than representing a burden.

As Robert Repetto of the World Resources Institute (WRI) and his colleagues have noted: "A country could exhaust its mineral resources, cut down its forests, erode its soils, pollute its aquifers, and hunt its wildlife and fisheries to extinction, but measured income would not be affected as these assets disappeared. Ironically, low-income countries, which are typically most dependent on natural resources for employment, revenues, and foreign exchange earnings are instructed to use a system for national accounting and macroeconomic analysis that almost completely ignores their principal assets."[6]

From an environmental perspective, research has been done to try to learn how great an effect these omissions and limitations of GNP have on politics and economics. Repetto and a team at WRI recalculated Indonesia's national income accounts with a procedure that included certain environmental considerations, and found the results to be significantly different from the official ones. While the reported gross domestic product increased at an average of 7.1 percent between 1971 and 1984, the WRI estimate of "net" domestic product rose by only 4.0 percent a year. And their calculations included only petroleum, timber, and soils on Java; they did not even consider such other important resources as natural gas, coal, copper, tin, and nickel, or non-timber forest products or fisheries. Had they been included, the differences would have been even greater.[7]

In testimony before a U.S. Congressional committee in 1989, Repetto explained the bottom-line importance of undertaking such revisions: "If depletion of natural resources can no longer masquerade as income growth, governments tempted to engage in environmental deficit financing will be less able to hide behind a reassuring screen of economic indicators. Policies that promote destructive and wasteful uses of natural resources will no longer be justified so easily as necessary for economic growth."[8]

A number of countries have begun work that may eventually lead to improved national resource accounting. Australia, Canada, France, the Netherlands, and Norway have programs to compile accounts on natural resource stocks and stock changes. Major changes, however, appear unlikely in the near future. The United Nations Statistical Commission has a System of National Accounts that many countries use. The Commission has been under pressure to update that system, but it has decided not to make any fundamental changes, although in 1993 it will at least include guidelines for incorporating environmental damages in national accounts.[9] The basic system is reviewed every 20 years, so it will be 2011 before the next review session is completed.[10]

Aside from the institutional and political inertia against changing the income accounts, another difficulty exists. Few good comprehensive statistics exist on environmental depreciation or its social counterparts. Many countries do not know accurately the state of their soils or their rivers. And methods are only just becoming available for integrating existing data into the numerical columns of the income accounts. Statisticians struggle just to maintain today's accounts and do not feel prepared to master a new system.[11]

Nevertheless, the resources are available to improve both the statistics and the income accounts. One of the purposes of the WRI study was to demonstrate that a small number of people with limited resources could shed substantial new light on the situation in a country that previously lacked the information to make sustainable policy decisions.

Moreover, even improvements that fall short of changing GNP calculations can still cause constructive change. For example, many indexes calculate environmental or social elements of national welfare, and they can be used to augment available information. In combination, they offer a more complete picture. The United Nations Development Programme (UNDP) has created the Human Development Index (HDI), which is an aggregate of longevity, knowledge, and the command over resources needed for a decent life. UNDP uses life expectancy at birth, adult literacy and average number of years of schooling, and gross domestic product per person adjusted for purchasing power to calculate its index, which runs on a scale from 0 to 1. Those three elements may measure a variety of underlying factors. For example, a high average life expectancy indicates good common access to health care, good nutrition, and access to clean water.[12]

Measured by the HDI, countries have a different ranking than when they are listed by GNP. Canada and Sweden, for example, come out near the top of the list, ahead of countries with higher per capita GNPs such as

Switzerland and Norway. And rankings are changing as the HDI is further developed. As more data become available, for example, the HDI will include other elements of socio-economic development. Information on sex inequalities applied to the HDI caused Japan to fall from the number one ranking to number 17, while Finland, ranked number 13, moved up to take the top spot.[13]

Another new measure, put forth by Herman Daly and John Cobb, is the Index of Sustainable Economic Welfare (ISEW). This is more comprehensive than the HDI, including environmental degradation, distribution of consumption, and other factors. It has only been calculated for the United States, and the data it needs will be difficult to obtain for many countries. Nevertheless, it is a positive step on the way to decision making based on realities rather than incomplete information. The ISEW calculations show an increase in welfare in the United States from 1950 to 1976, paralleling the rise in GNP, but from there on it shows a decline, with a 12-percent drop in 1988. Just looking at the GNP creates a much more optimistic view of national welfare and future potential.[14]

Failures of Economic Policy

Although economists beginning with Adam Smith a century and a half ago have talked of an "invisible hand" that governs economies optimally, no such hand exists. Instead, economics is full of both distortions that make it subservient to political ends and failures that prevent it from achieving those goals.

Chief among them may be subsidies—payments by a government to protect, encourage, or change a particular industry. Farm export subsidies, for example, are stipends from a government that enable exporters to eliminate or dump surplus commodities that are not absorbed by domestic markets.

The effect of these subsidies is complex. They may help protect American farmers and preserve their way of life and the culture that surrounds it. Some say they also improve American food security and keep balance in the economy. On the other hand, their cost to the U.S. Treasury is spectacular—$13.9 billion in 1989.[15]

The effects of U.S. export subsidies may be even larger in other countries. Agricultural crops, for example, are exactly what the poorest countries depend on for income from international markets for their development and survival. But some farm programs result in overproduction of crops in the United States, and when those crops enter international markets, the additional export subsidies drive down prices even further and keep them well below the costs of production. In Africa, these deteriorating terms of trade have translated into losses equivalent to 10 percent of the gross domestic product in recent years.[16] Fluctuating and low commodity prices may be the most serious barrier to international development, and they highlight the connections between the domestic policies of one country and the economic, social, and environmental opportunities of others.

Many of the worst environmental abuses are occurring through ongoing government subsidization because of vested interest groups. A lot of the environmental organizations working on the farm bill argue that the current system of subsidies is destroying the small farm. It is mainly the big agricultural enterprises that benefit from these subsidies because the model of agriculture being promoted is so capital-intensive that there is no way the small farmer can survive.

The whole movement toward sustainable agriculture, biological, organic agriculture, which a decade ago was more of a fad, now is a major economic phenomenon. The National Academy of Sciences came out with an important study recently on this, a fact that is immensely important internationally, because the U.S. model of agriculture has been promoted all over the world in developing countries, and is promoted through development banks.

Only now has a strong indigenous economic and social movement of family farmers emerged. I think that it is worth looking at as an example of how there could be an economic and social evolution within the U.S. (that might already be occurring) toward a more sustainable system.

Bruce Rich, *United States*
Environmental Defense Fund
Earth Summit, p. 144

Beyond the subsidies that support agricultural incomes, other forms of support promote certain techniques of farming over other methods. In Indonesia during the early 1980s, subsidies made pesticides available to farmers at as much as 82 percent less than their market value—at a cost to the government of $100 million every year.[17] And in Egypt at that time, pesticide subsidies of 83 percent of retail value cost more than $200 million a year. Worldwatch Institute reports that the Egyptian government spent more on pesticide subsidies in 1982 than it currently spends on health.[18]

In Brazil, subsidies for cattle between 1965 and 1983 cost $1.4 billion, and the effect has been dramatic deforestation and a cattle ranching boon that left poor Brazilians more dependent than ever on land and capital owned by a tiny elite.[19] Another subsidy, in the form of income tax credits of up to 50 percent, was available to those who invested the resulting savings in the development of the Amazon region. The suspension of that allowance in 1988 is credited with part of the decline in deforestation beginning at that same time.[20]

The governments of many countries, including the United States, partially cover the costs of high-technology farm machinery as well as chemical fertilizers and pesticides. This encourages farmers to rely on them rather than more sustainable and balanced methods of farming that do not involve elaborate inputs. It also tends to increase the amounts of chemicals present in the food that Americans eat and to make farms more dependent on both government finances and bank loans to pay for expensive chemicals and machines. (See the section on Agricultural Degradation and Disruption in Chapter 5).

Previous American subsidies that promoted soil erosion have recently been partially reversed by the U.S. Conservation Reserve Program, under which farmers who plant their most erodible land with trees or grass for 10 years receive about $120 per hectare every year from the government. They suddenly have an incentive to protect the long-term value of their soil and replace the nutrients that heavy chemical-based farming drew away. By 1990 almost 14 million hectares of land were under the program, and soil erosion had been reduced by more than one third, from 1.6 billion tons to 1.0 billion.[21]

Similarly, new programs in several states are now attempting to make utility regulation more oriented toward energy efficiency and pollution reduction. Under most current regulations, utility profits rise lockstep with electricity sales, and so companies have little incentive to encourage the use of the most efficient technologies. Even though improvements in efficiency increase the availability of energy at far less cost than supplying more would, the benefits enjoyed by utility companies do not usually increase with efficiency.

In California, the Public Utilities Commission approved a proposal in 1990 by the three largest electric utilities to tie earnings to energy savings. One company will be allowed to charge electricity rates that yield an annual return of 14.6 percent on its conservation investments, which is significantly higher that the rate of 10.7 percent it would get from investment in new production facilities. The other two utilities will receive profits amounting to 15-17 percent of the energy savings they achieve. Together, the efficiency programs will cost an estimated $500 million over two years, but may save more than twice that amount in reduced power needs.[22]

Other failures of the market revolve around the difficulty of putting a numerical value on products and resources, as discussed earlier in this chapter. International trade, for example, is the way we allocate natural resources worldwide. In theory, it is intended to achieve the optimal use of those resources by following the mechanisms of the free market. And that can work, but only to the extent that the market is able to account for the value of the resources it distributes.

If the market judges the worth of a resource to be lower than the full value that the resource offers to society, the market will not treat it optimally. If the market price for a forest includes only its short-term value but ignores the long-term and interconnected ones, then the market receives the wrong signal. Such misallocations of resources are a major barrier to sustainable development, as they take into account only limited time frames and partial functions of the natural and human environments. (See Chapter 1 and the section on Improper Resource Accounting in this chapter for more on environmental accounting.)

Similar misuse of resources occurs in corporate investment. If a corporation will benefit from using a river, or the atmosphere, for free, it will have no incentive to use that resource optimally. It is only when the full costs are factored into business considerations that decisions will be directed toward the fullest benefits. If businesses make decisions based on short-term profits they will have less reason to invest in long-term effectiveness through education programs for their employees, efficiency programs for their technology, and environmental and public health.

Coming full circle, attempts to maximize short-term gain at the expense of the longer term can also undermine the ability of free market mechanisms themselves to meet society's goals. For example, calls by American businesses for protectionist policies that inhibit international trade make it easier on them in the short term and may raise this year's profits, but they eliminate natural market-based incentives for innovation and improvements in efficiency that would otherwise work to meet both goals of long-term corporate profit and environmental protection. By eliminating international competition in trade, U.S. companies choose to grow soft rather than develop tougher and tougher standards of conduct regarding efficiency.

Even when market mechanisms succeed in generating production to meet society's needs, they often fail at distributing those products. The benefits often go to a small segment of the population, while the major part of society has inadequate opportunities for education, participation in economic systems, good health, and a better quality of life. Such a situation is not in the interests of sound future economics, and the market is not succeeding in any fundamental sense.

In the countries that have the most uneven distribution of assets, national debt is often large, social and environmental programs are often weak, and investment in those countries is often sporadic or small. In Brazil in 1982, the richest 20 percent of the population held 64 percent of the national income, while the poorest 20 percent controlled just over 2 percent.[23] The average income of the richest one fifth was 28 times greater than the average income of the poorest. Around that same time, Brazil also had to pay about $1 billion every month in interest and repayments on debt.[24]

In the United States, the richest 20 percent of Americans controlled nearly 45 percent of the national income in 1991, while the poorest 20 percent held just under 5 percent.[25] The average income of the top one fifth was almost 10 times greater than that of the poorest 20 percent of Americans.[26] Like Brazil, the United States has a huge national debt and an uneven distribution of social well-being.

In economics, as in other areas like food availability, distribution can mean everything. Latin American cattle ranchers own a disproportionate amount of the national assets, and they have often turned it to unsustainable uses that generated large incomes for themselves at the expense of the land they possessed and the people who work for them. In some countries, land reform has proved to be one of the most effective means of pursuing a more equitable distribution of income. In both Taiwan and South Korea, massive land reforms in the early 1950s helped narrow the gap between rural and urban incomes and provided the solid basis they needed for economic growth in the 1960s.[27]

Small landowners who produce for their families and their children have a high motivation for valuing the future, while large holders who live far from their land have little incentive to do so. Smallholders spend their income locally and strengthen community structures, including those for health facilities, local government, and schools. Large landowners invest abroad, contributing to capital flight and reinforcing the lopsided nature of their economies.

Market principles do not necessarily reinforce the types of integrated rural development that have the most potential to reach many poor people, allow them to manage the environment wisely, and develop a balanced economy. They often funnel wealth and economic opportunity away from women—the primary food producers in most countries and the ones who often have the greatest responsibility for taking care of families, education, local environments, and health. Women often do not control scarce resources, credit, and services. Yet economic incentives revolve around the control of those inputs, and sustainable development depends on them deeply.

As long as land and productive capabilities are unevenly distributed, policy changes meant to protect the resource base can instead lead to greater inequities of ownership. The poor are shut off from the inputs and services needed to take advantage of new programs, and the advantages go to the rich. Instead, the poor are forced to overplow fragile land, overplant depleted soils, overfish local bays, and overuse other exhausted natural resources in order to survive.[28]

Distribution of goods varies not only geographically but also over time. The unequal production and consumption of food worldwide is mirrored by the shortages that sometimes afflict people who otherwise have enough to eat. Market forces are not always equipped to compensate for temporary food shortages, and they must be augmented by other means of protecting people from famine. They must also do so without undermining the systems of production used in those areas normally, because food aid can disrupt local economies by bankrupting farmers who cannot compete with incoming supplies. Food production should be shifted away from over-production in industrial market economies and toward production in developing countries. And there, food must be shared equitably. Improvement in these areas will pay for itself many times over in economic, social, and human terms.

Military Spending

For the last 14 years, Ruth Leger Sivard has published a valuable compendium of data called *World Military and Social Expenditures.* In the most recent edition, she notes that during the 1980s "global expenditures on arms and armies approached one trillion dollars a year—$2,000,000 a minute. The number of wars under way reached an all-time peak; three-fourths of the people killed in them were civilians."[29] It is easy to see the effect this has on sustainable development: it not only drains financial resources but also occupies peoples' time with destructive activities and causes devastating injury and loss of life.

Moreover, large militaries can actually reduce security by increasing tension and distrust. Military-industrial complexes can take on lives of their own and carry an economy farther than national priorities would have it go, as President Eisenhower pointed out some 35 years ago. Such spending does nothing to resolve the underlying problems that caused conflict in the first place.

And it diverts resources away from the areas that have the potential to reduce conflict by improving the quality of life and well-being of people. (See Figure 4-1.) Those who are well off are substantially less likely to be involved in a war than those whose fundamental needs go unmet. Similarly, those whose human rights are protected and who have access to political and economic participation are much less prone to engage in wars. No two democracies have ever fought each other.[30]

In 1988, the world spent $36,000 per soldier but just about $1,100 per student. "The world now has 26,000,000 people in the regular armed forces, another 40,000,000 in military reserves, a stockpile of 51,000 nuclear weapons, 66 countries in the business of peddling arms, 64 national governments under some form of military control, and 16 wars under way," according to Sivard.[31]

The United States ranks first among countries in military expenditure, but only eighth in per capita public expenditure for education, thirteenth in maternal mortality rate, fourteenth in per capita public expenditure for health, fifteenth in life expectancy, twentieth in percent of population with

FIGURE 4-1: *Military Spending and Human Development Performance*

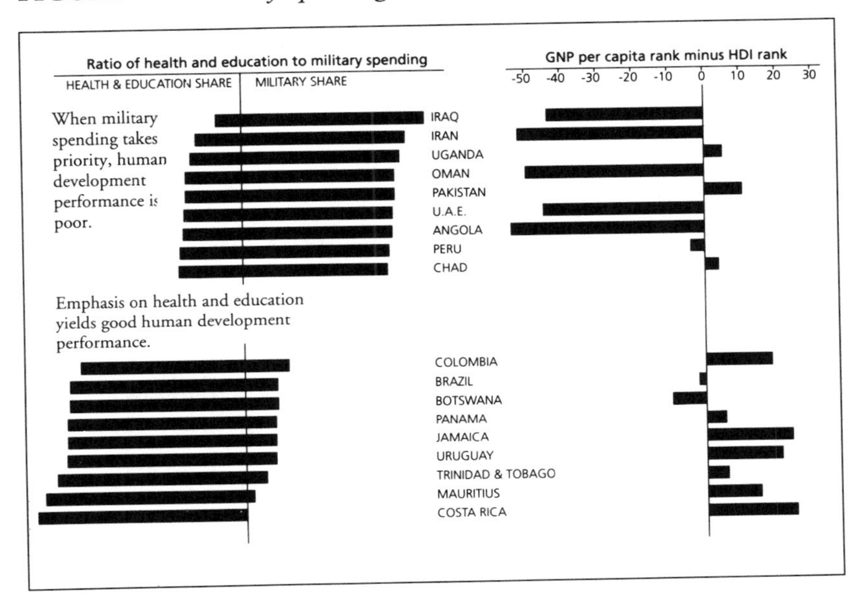

SOURCE: UNDP, *Human Development Report 1991*, p.83.

access to sanitation, and thirty-sixth in percent of infants with low birth weight.[32] The increase in the annual U.S. defense budget from 1980 to 1991 was $164 billion, more than the $146 billion rise in the nation's overall budget deficit.

By leaving the United States in debt and operating below its potential of providing health and education spending for its people, military spending detracts from exactly the areas at the heart of sustainable development. Corporate and government research that could go toward productive enterprises instead goes toward arms. Instead of investing in people, the nation invests in weapons.

Environmental problems that go unsolved increase and can contribute to conflict. Rivers, the atmosphere, and ecosystems cross national and state borders, and when they are degraded they contribute to tensions among the countries and states that share them. Similarly, social problems that go unsolved grow in dimension and can cause unrest both inside and outside national borders. The existence of "environmental refugees" ties them both together, as the impoverished victims of environmental degradation cross national borders and affect the security of neighboring states.

Directing government spending toward environmental and social goals may make a nation more secure than spending on the military does, as Costa Rica has discovered. The government there abolished the army in 1948 (although it did recently establish a paramilitary civil guard).[33] At the same time, considerable effort was put into improving health and education. The nation has among the highest literacy and lowest infant mortality rates in Latin America, its fertility rate has been cut in half over 25 years, and its maternal mortality rate is only three times that in the United States—and one fifth the rate in Colombia and less than one tenth that in Peru.[34] The U.N. Development Programme (UNDP) puts Costa Rica in the highest category in its Human Development Index, above all other Latin American countries except Chile, Uruguay, and a few island nations.[35]

The same principle applies in the United States. For an individual American, the threats of cancer, heart disease, or a car accident surpass the threat of invasion by a foreign military. Spending on cancer research, hospital care, public health in general, and automobile and highway safety has a much greater impact on the security of Americans than does military spending.

In many areas of the United States and abroad, significant threats exist from violent crime most often carried out by people suffering from drug addiction, desperation, and poverty. Spending on inner-city jobs programs, training programs, and small inner-city businesses, or at least more adequate spending on American police forces, would improve the situation. Military spending will not.

The money to fund those programs can be found. At this juncture in human history, with the cold war over and former enemies scrambling to help each other, the place to find it is clear. UNDP notes that military spending in industrial countries could be cut 3-4 percent a year during the 1990s, yielding $1 trillion for urgent domestic social problems and enough funds to double development assistance to the South. And if developing countries freeze their own military expenditures, rather than doubling them as expected, they would save $15 billion a year that could be put to better use—funding sustainable development.[36]

Debt and Structural Adjustment

The debt crisis is a disaster for more than just national economies and the international economic system. Countries struggling to meet payments on the interest on their national debts find it hard to invest in environmental protection and in the health, education, and productivity of their people.

In turn, this lack of investment in the future makes it that much more difficult for debtor nations to overcome the weight of their debts and the interest they owe. The increase in productivity and effectiveness at solving problems that was intended to result from the loans was meant to enable their repayment. But that increase has been replaced by stagnation, poverty, and the hopelessness that accompanies them. It is a downward cycle instead of an upward one.

Developing countries now pay some $35 billion a year more to industrial countries to meet interest and repayment of loans than they receive in the form of development assistance. This marks a net flow of money away from the developing world to the industrial one.[37]

Such a negative flow of financial resources is a human, environmental, developmental, and economic threat. It undermines both the present and the future potential of people to provide for themselves and preserve for the future. Instead of using the soil, forests, plants and animals, and cities that support them in sustainable ways, poor people in debtor nations desperately overuse whatever resource is at hand to pull out another day of survival for their families.

Fields that will produce food for many years if properly fertilized and planted with rotated crops are instead farmed overintensively and quickly depleted of nutrients. Trees that anchor soil to the earth, store water, fertilize fields, and balance ecosystems are cut in a desperate effort to heat and light homes by people who have no other choice. Squatters' settlements spring up around the cities of the poorest countries as people make a last-ditch effort to find livelihoods by migrating to the city.

While debt is not by any means the only cause of such socially and environmentally destructive desperation and poverty, it has become a major aggravant. And it intensifies many of the other causes by redirecting government and economic support away from the world's poorest people. It has recently been joined by another force that redirects resources away from the human and environmental foundations of development and sustainability—conditionality on new loans.

When a nation cannot meet its debt payments, it must apply for help to an international institution and take a transitional loan to pay the interest on its previous ones. The organization that provides those holdover loans is the International Monetary Fund (IMF) in Washington, D.C., the sister organization to the World Bank. IMF loans, however, come with conditionality strings attached—a set of actions that borrower countries must agree to in order to receive the loan, even though those countries have little choice but to accept the "structural adjustment" programs.

These programs are also called "austerity programs" by people less sympathetic with the IMF plans. Basically, they set out to do two things: reduce expenditures and increase revenues. To cut expenditures, they eliminate supports on staple foods, fuel, transportation, utilities, and government programs for health care, education, environmental protection, and other vital goals. To raise revenue, they promote overintensive production of crops for sale to bring in foreign currency rather than a sustainable balance of field rotation and production of crops for local consumption. They encourage such policies as the cutting of forests for timber exports, but ignore the larger value of sustainable forest use, including conservation. And since many debtor countries produce similar goods for export, world markets are flooded with those products and their prices fluctuate dangerously, thereby undermining the intended income from their sale.

Until progress is made on alleviating the debt burden many developing countries face, much of which would never be paid back anyway, governments will not be able to invest in the many programs needed for a sustainable future. And the stress placed on environmental systems, on people's lives, and on their social and cultural systems will continue to rise.

CHAPTER 5

Environmental Systems

The economic systems described in Chapter 4—combined with the unrelenting growth of world population, the inefficient way energy and other resources are used, and the wasteful consumption patterns in industrial nations—have put enormous stress on the natural life-support systems on which all economic and human development depends.

Biodiversity and Biotechnology

The term biodiversity encapsulates an explosion of colors, shapes, textures, sounds, motions, and prehistoric activities. All the patterns of a rain forest or a coral reef are contained in the word biodiversity. Prehistory still lives in the genetic structures of plants and animals in the most remote and untouched parts of the world, organisms that appear much as they did millennia ago. The designs of the future exist there as well, in shapes and forms that science has yet to imitate.

All life depends on other lives: Survival moves up the food chain, from the tiniest microorganisms to the hugest trees and the hungriest humans. Habitats evolve, touched irreversibly by the interactions of plants, animals, and bacteria. Every part of the biosphere provides a function for whatever creature can use it most effectively or ingeniously. Balance is maintained through the interplay of innumerable species acting in innumerable patterns.

The loss of multiple species can begin with the degradation of just one. Each species makes a connecting link on the food chain, and each plays a role in such processes as decomposition, pollination, and the transport of nutrients. When those chain links are broken, whole ecosystems suffer. When structural changes, such as the mass removal of timber, overtake ecosystems, entire habitats grind immediately to a halt, and almost all their organisms disappear.

Damage to parts of an ecosystem can proceed to a certain point without destroying the system as a whole. Once that threshold is reached, however, the collapse of an entire ecosystem can follow the degradation of what might appear to be a specific or individual resource. In *Extinction,* Paul and Anne Ehrlich liken the situation to someone removing a rivet from the wing of an airplane. Taking just one rivet is unlikely to cause a crash since airplanes, like ecosystems, have designs that include considerable redundancy. But continuing to take one after another will sooner or later lead to sudden disaster. The removal of one last critical rivet will cause a failure of the system's integrity. The plane will crash.

The removal of the critical rivet is, in ecological terms, the crossing of a threshold. There are many such thresholds in nature. Removing species from ecosystems can cause those systems to unravel. Taking out all the trees in an area can alter a watershed, which in turn can trigger a loss of topsoil, the loss of more trees and then soil, and so on.

The losses from ecosystem disruption and habitat degradation translate into losses in other regions. Ecosystems play vital roles in pest and flood control, weather patterns, absorption of waste materials, and other life-sustaining services. They play global roles in the absorption of climate-altering gases, the production of oxygen, maintenance of balanced sea levels, and protection from solar radiation. They offer resources and "environmental machinery" to industry. Degraded ecosystems fail to accomplish any of these roles.

Healthy natural ecosystems have aesthetic significance, receive millions of tourists every year, and serve as symbols of human values. The biodiversity that they contain supplies industry, science, and medicine with the genetic bases for what later become new pharmaceuticals, plastics, textiles, crop varieties, and countless other industrial innovations, scientific discoveries, and medical solutions.

In agriculture, for example, more than 90 percent of U.S. food production is derived from species that were introduced into the country, while crop breeding programs that draw on genetic diversity add an estimated $1 billion a year to the value of our agricultural production.[1] And in health care, about four-fifths of the people in developing countries still rely on traditional medicines.[2] Even in the United States, more than 40 percent of all prescriptions depend on natural sources.[3]

Despite the many known services provided by a biodiverse world, scientists do not even know how many species exist. Many assume the total is 10 million, but it could be 30 million or maybe 50 million. Only 1.4 million species have been identified and recorded. Whatever the total is, we know that biodiversity within the global system is being reduced as environmental systems have come under stress. Without human intervention, 1 to 10 species a year would disappear in the natural evolution of life. Yet conservative estimates are that 100 species become extinct now each and every day. Current extinction rates among birds and mammals are perhaps 100 to 1,000 times what they would be in unperturbed nature.[4]

Biotechnology, or the transfer of genes from one organism—a plant or animal or even a microorganism—into another is a science and industry that is growing dramatically in the United States, Canada, Europe, and Japan. Companies are producing food additives and new products like drugs and cow hormones that will dramatically alter the way we live. It is not clear how safe are most of these products or whether the "producers" are getting what they deserve for them. Biotechnology products, which come from vast stores of genetic resources like coral reefs and forests are "technically" borrowed from their "producers," farmers and indigenous peoples, who are not paid for their contribution. Part of any resolution or convention on biological diversity must include some form of compensation for these gene keepers.

Edward O. Wilson of Harvard University and Paul Ehrlich of Stanford recently concluded that at least a quarter of all species could be exterminated within 50 years if tropical rain forests continue to disappear at the current rate.[5] Deforestation in the tropics alone destroys daily an area as large as the Redwood National Park in California and yearly an area as large as Washington state. But the problem is not confined to the tropics. The United States has lost 54 percent of its wetlands since colonial times,[6] and the tallgrass prairies that once covered central North America have been reduced in area by fully 99 percent.[7]

In the United States, the species threatened with extinction vary in different parts of the country. Half the nation's endangered plant species are found in California, Hawaii, and Puerto Rico; indeed, 10 percent of Hawaii's native flora has already disappeared, and 40 percent is threatened. Some 80 percent of the endangered and threatened fish are in the arid Southwest and the southeastern part of the country.[8] Even protecting animals in national parks cannot ensure their survival, as Table 5-1 indicates.

By early in the twenty-first century, more than a million species may become extinct. This would be a more massive loss than during any period of geological history, including the extinction of the dinosaurs and the ice ages. It is a problem for much more than just the countries where the greatest losses of biodiversity are taking place. It is a problem for all regions of the planet, and one that will endure forever. It is a loss of life itself, in many of the forms it takes, and with all the supporting roles and interactions it performs for the biosphere as a whole.

The maintenance of biodiversity is deeply embedded in the concept of sustainable development. Conserving human and natural options for the future depends on it. The cyclical processes of nature can continue, or they can end in a one-time activity. It all depends on what stresses are placed on the present systems. The 1980s were a time of enormous changes in biodiversity. The 1990s must be the time when we stop both wasting our present resources and throwing away those of the future.

Forests

In many ways, the loss of forests can be traced back to human systems of education, government, and natural resource accounting. Economics textbooks that teach how to value natural resources do not discuss the long-term ecological value of living trees and the roles they play. They often ignore the value of the other crops supported by the environmental services provided by intact forests, as well as such tangible and intangible uses as sites of ecotourism and aesthetic enrichment. Governments similarly provide incentives to cut down forests based solely on the short-term bottom line. International trade allocates timber based on where it can be obtained most cheaply and efficiently, but not on where it would leave the most continued potential for long-term benefit.

Just as forest ecosystems evolved over millennia through interactions

TABLE 5-1: *Habitat Area and Loss of Large Animal Species in North American National Parks, 1986*

Park	Area (square kilometers)	Original Species Lost (percent)
Bryce Canyon	144	36
Lassen Volcano	426	43
Zion	588	36
Crater Lake	641	31
Mount Rainier	976	32
Rocky Mountain	1,049	31
Yosemite	2.083	25
Sequoia-Kings Canyon	3,389	23
Glacier-Waterton	4,627	7
Grand Teton-Yellowstone	10,328	4
Kootenay-Banff-Jasper-Yoho	20,736	0

SOURCE: Worldwatch Institute. Based on William D. Newmark, "A Land-Bridge Island Perspective on Mammalian Extinctions in Western North American Parks," *Nature,* January 29, 1987.

with the animals and other organisms of the forest, they are now evolving rapidly through interactions with people and human economies. People consuming products in one region—including meats, vegetables, and oil-based products—and decision makers living in one area now contribute to the evolution of forests in distant places. Much as deforestation contributes to increased soil erosion and harm to the ecology of other regions, human activities in one part of the world now contribute to increased rates of change of forests in other regions—demonstrating once again the complex connections binding our actions with the health of the planet we depend on.

In cases where economic valuation systems fail to measure the full worth of resources, it is politics and environmental policy that can step in to change national and international priorities. Environmental legislation can set aside parks that will last into the future, and can use tax structures and other economic incentives to minimize the production of waste, maximize efficiency, and reduce the use of harmful chemicals.

What is lacking is a plan. The U.S. government already balances economic and environmental priorities, but it does so without an overall plan for making the two mutually compatible. Instead, it follows a piecemeal approach, where an environmental issue such as threats to the spotted owl in the Pacific Northwest is dealt with as it arises, but is not addressed before pressures have mounted to the point of stress. Moreover, solutions tend toward linear patterns and one-time actions, while the problems they must address work in cyclical and interconnected ways. Instead of making natural and human patterns compatible, today's solutions tend to patch injuries and continue with business as usual.

Rather than presenting obstacles for each other, economic and ecological systems must contribute to solving each other's problems. Economic actions that degrade forests work counter to economic goals: They undermine the livelihoods of indigenous peoples who live there, they remove an absorber of carbon dioxide, and they threaten other natural resources needed for economic activity, including watersheds and coastal ecology.

It is in this context that efforts to concur on a global forest agreement must be seen. The multiple roles of forests in all areas must be considered. Halting forest degradation and deforestation should go hand in hand with efforts to stop clear-cutting in industrial countries and to reforest areas that were carelessly stripped of trees in the earlier push for industrialization and

---

Like an autoimune disease, in which a body's own defense system attacks healthy tissue, our economy is assaulting the very life-support systems that keep it functioning. A fundamental restructuring of the rules and practices that shape economic activity is needed to stem this self-destruction.

Lester R. Brown et al., *United States*
Worldwatch Institute
Saving the Planet, p. 115

---

economic development. Sermons to the Third World about the need to stop deforestation have a hollow ring coming from nations that followed a similar path scant decades ago.

The many vital roles played by ecosystems are abundantly evident in the world's forests. *The Global Ecology Handbook* describes the importance of tropical forests in particular: "Forests moderate air temperature, maintain the hydrologic cycle by absorbing rainfall and releasing moisture to the atmosphere, and take in carbon dioxide and generate oxygen through photosynthesis. They recycle nutrients and wastes, control soil erosion and sedimentation of waterways, and regulate stream and river flows, helping to moderate floods and droughts. Tropical forests also prevent or limit landslides and rockfalls during rainstorms and earthquakes, and moderate damage from tropical cyclones."[9]

Beyond these key environmental services, forests of course are harvested both for the plants, fruits, and other growing crops they offer and for the trees that constitute them. Trees can be taken from forests in two ways: by extractive methods that do not harm the forest ecosystem and allow it to produce new crops every year, or by clear-cutting, which takes an entire region of forest. That region can then be replanted as a monoculture or with mixed native tree species, or it can be left barren. Either way, the original ecosystem cannot be restored for thousands of years, as it is a product of the evolution of a complex, integrated system made of countless microorganisms, plants, and various stages of tree growth.

New technologies for logging have allowed great success in some areas in the nondestructive extraction of trees for timber. And studies have demonstrated that renewable use of a forest for the year-after-year harvesting of crops produces more profit than a one-time clearing. Efforts are now being made, particularly in Latin America, to set up "extractive reserves" that keep forests intact but allow local people to harvest fruits, nuts, rubber, and other products for local use and for export. In the United States, edible evidence of this new approach is available in ice cream that uses Brazil nuts harvested sustainably in the Amazon.

Unfortunately, clear-cutting is still the norm today. Worse, the damage extends beyond the area harvested. In the U.S. Pacific Northwest, for each 10-hectare virgin forest that is clear-cut another 14 hectares are degraded from exposure to wind, nonindigenous species, and local changes in climate.[10] The amount of forestry that is done in a sustainable manner in the tropics is pitifully small according to a recent study. (See Box 5-1.)

Box 5-1: *Sustainable Tropical Forestry?*

In 1988, the International Tropical Timber Organization (ITTO), which is devoted to protecting the future of the tropical timber trade, enhancing nontimber products and services, and conserving forest ecosystems, commissioned a worldwide survey of how much forest was under sustainable management. The survey found only a negligible amount of forest being managed for sustainable, long-term timber production—4.4 million hectares out of a global total of 828 million hectares of productive tropical forest. In other words, timber is produced sustainably on less than 1 percent of the exploitable tropical forests. (Sustainable timber production is predicated upon doing nothing to reduce irreversibly the forest's potential to produce marketable timber.)

While searching each tropical region for sustainably managed forests, ITTO researchers painted a discouraging picture. In Latin America and the Caribbean, they reported, the total area being sustainably managed at an operational level is limited to 75,000 hectares in Trinidad and Tobago, of which 16,000 have been declared as fully regenerated after logging. They found similar conditions in Africa. Asia was different in that all the forests under logging concessions are nominally under management. However, [ITTO noted], "there is a very great difference between theory and practice." Tropical timber harvesting has the longest history in Asia, and examples of sustained timber management include the Mae Poong forest in Thailand and peninsular Malaysia's selective management system.

SOURCE: Excerpted from World Resources Institute, *World Resources 1990-91* (New York: Oxford University Press, 1990), p. 106.

Primary forests—those not altered irreversibly by human activities—are the richest pools of biodiversity. Worldwide, these stands are now only about one fourth the size they were before settled agriculture began.[11] The United States has just 15 percent of its primary forests intact, but if Alaska is excluded, the figure drops to less than 5 percent.[12] Tropical rain forests—the richest in biodiversity of all primary stands—are disappearing worldwide, with about half the original forests already gone. Debate about the current extent of this hinges largely on what is happening in Brazil, where deforestation appears to have peaked in 1987 and declined somewhat recently due to government policies and wetter weather.[13] Still, it seems that 17 million hectares of original tropical forest is disappearing annually.[14]

One result of this massive deforestation is a worsening of global warming, as the release of carbon dioxide during all this tree cutting accounts for 15-30 percent of annual carbon dioxide emissions.[15] Another result of shortsighted and unsustainable practices is a threat to the very thing developing countries rely on logging for now: foreign exchange. The World Bank expects that the number of tropical wood-exporting countries will drop from 33 to just 10 during the 1990s due to overcutting of forests.[16]

A third result is the disruption of people's lives—and livelihoods. For some 50 million indigenous people, rain forests are home and a source of life. (See also Chapter 6.) As trees are felled, all that changes. In Venezuela, for example, more than 30 percent of the Yanomami population has died from measles and whooping cough brought in by outside workers. As a Kayapo leader in Brazil explains: "We are fighting to defend the forest. It is because the forest is what makes us, and what makes our hearts go. Because without the forest we won't be able to breathe and our hearts will stop and we will die."[17]

Oceans and Fisheries

After rain forests, the oceans that cover two thirds of the earth's surface support the most biodiversity of any habitat. Coastal areas within 200 miles of land are the most productive of all, and sustain more than half the ocean's life and nearly all the world's fish catch. In these areas, nutrients wash down from the land, surface winds and ocean currents dredge up nutrient-rich sediments from the sea bottom, and sunlight promotes plant growth on the shallow seafloor. Coastal areas are also home to most human beings—by the year 2000, an estimated three quarters of all people will live there.[18]

On the continental shelf of the eastern United States, one of the world's richest fishing areas, at least three quarters of commercially valuable fish spend part of their life in the regions where rivers carrying fertile silt meet the ocean. These estuaries support a long and elaborate chain of life, from protozoa to fish-eating mammals, and they provide essential habitats and breeding grounds for a wide range of wildlife.

Salt marshes and mangroves, two other types of ecosystems located where sea meets land, play integral roles in life cycles. They are associated with the growth of offshore seagrasses that provide food for ducks and geese, sea turtles, and aquatic mammals. They also trap nutrients, filter out pollution, and prevent coastal erosion. Mangroves are found in the tropics, and are probably the most productive coastal ecosystem.

Coral reefs are diverse, intricate, and delicate. They require clear water, bright light, constant amounts of salt content in the water, and a warm temperature. They support a spectacular array of life. But they are easily disturbed by environmental changes. Soil erosion and the dumping of waste make water opaque and block sunlight. Coral mining, blast fishing, overfishing, and collection of coral all harm the reefs, which are among the slowest-growing of all ecosystems.

Island ecosystems, many of which evolved in the absence of human influences, are particularly vulnerable to human impacts. Seabirds and sea turtles, oceanic marvels tethered for part of the year to their breeding sites are the most obvious early warning system of stress. Disturbance caused by the presence of people; habitat destruction caused by construction or by introduced animals such as goats and rabbits; and introduced predators such as rats, cats and snakes have eliminated seabirds and sea turtles from thousands of islands. The terrestrial animals and plant assemblages of islands are less obvious victims. One well-known example is Hawaii, where human impacts in the past 200 years have caused the extinction of about a fourth of the islands landbird species.

Ecosystems of coastal areas are changing faster than at any time in history. These habitats for nature and people are touched, and sometimes overwhelmed, by municipal sewage, industrial and oil and gas waste, urban, agricultural runoff, and fallout of atmospheric pollutants. With about 35 percent of U.S. sewage ending up in marine waters, the problem of contaminated coasts is one that Americans are probably most familiar with.[19] In 1989, more than 40 beaches in New York, New Jersey, and Connecticut were closed for this reason during the peak 26-day vacation period. Although Congress set a goal of zero discharges in the Clean Water Act of 1972, pollutants still reach coastal waters from some 20,000 industrial and municipal sources.[20] Ninety percent of material currently dumped in U.S. waters is dredged from harbors and channels.

Natural and artificial substances from pollution disrupt the balance of ocean systems. Among other changes, sewage and agricultural runoff introduce large quantities of nitrogen and phosphorus into coastal waters, which nourish algae, sometimes explosively. The algae deplete the oxygen in the water and suffocate other species—in areas that have come to be known as "dead zones." Algae themselves can also be toxic to marine life; between 1987 and 1989 as many as 3,000 dolphins may have died along the Atlantic Coast from eating contaminated fish.[21]

Heavy metals and synthetic organic contaminants from industry are major sources of freshwater pollution in all industrial countries. They tend to "bioaccumulate" by concentrating at higher levels of the food chain, leading to increased risk of cancers and reproductive abnormalities in fish, aquatic mammals, and humans. The U.S. regulatory system results in an elaborate and expensive pollution control effort by industry, but nevertheless sanctions by default the generation of certain amounts of waste that accumulate to environmentally unacceptable levels over time. Accidents, too, regularly strip all life from affected rivers. Continued reliance on toxic materials for industrial production will assure regular news stories about spills and fish kills.

Oyster harvests in the Chesapeake Bay have dropped by two thirds during the past 20 years, and commercial catches of striped bass (rockfish) fell from 6 million pounds in 1970 to 600,000 pounds in 1983. By the end of the 1980s, even recreational fishing of the Bay's premier sport fish was prohibited. Many attribute the area's declining marine life to this discharge of hazardous wastes. For example, an EPA study found more than 3,000 tons of toxic metals from industries in Maryland and Virginia enter the Chesapeake Bay each year.[22] All acknowledge the problems are complex.

Nonpoint source pollution runoff from streets, agriculture and other sources not originating in a pipe or other single location account for 34 percent of total coastal pollution in the United States.[23] Atmospheric contamination, such as the nitrates found in acid rain, is estimated to contribute 30 percent of the total amount of pollution found in marine waters worldwide.[24] By some accounts, more than 25 percent of the nitrogen entering the Chesapeake Bay comes from air pollution.[25] Yet these types of pollution are often out of sight and thus out of mind, since no photographable source offers impressive images. And oceans, with the accumulated wealth of life they support, are often outside of the public eye. Only a handful of celebrated species—such as whales—receive attention, while no comprehensive plan exists to preserve the habitat of millions of species.

Instead, unplanned or misplanned human actions reduce the long-term potentials of the oceans. Simple overfishing is an example. Fourteen years ago, the U.S. government declared a 200-mile boundary around its shores, banning most foreign trawlers that had fished there. It was an effort to save the fish crop from overexploitation, and it worked—temporarily. But American fishers soon went on a boat-building binge, encouraged in part by a federal loan program. They used new trawlers, with new fish-finding electronics, and lobbied hard to get bureaucrats to remove quotas on catches.

The fish catch went up, peaking in 1983. Since then, however, it has fallen sharply: Stocks are at record lows, and many New England fishers find themselves in dire straits. The *Wall Street Journal* called them "victims of a get-it-while-you-can mentality that could exhaust stocks beyond recovery." But the fishers have large mortgages now, on expensive boats, and cannot afford the luxury of not catching as much as they can as fast as they can, reports the *Journal*. Local people—many of whose parents and grandparents fished before them—are selling their boats and taking up other jobs, if they can find them, in what has become the loss of a traditional way of life that parallels the loss of an ecosystem.[26]

The world fish catch reached 97.4 million metric tons in 1988—perilously close to the 100 million tons that the U.N. Food and Agriculture Organization estimates is the annual sustainable yield.[27] Outputs in some major fisheries have levelled off or are declining, and the crash of the Peruvian anchovy fishery between 1970 and 1973 provided a sobering lesson about overfishing's effects. Closer to home, the harvest of Alaskan king crabs peaked in 1980 at some 84,000 metric tons, and then plummeted to below 7,000 tons in 1985.[28] So fishers the world over may soon find their lives and social systems under stress from the pressures put on this natural system. Whether it blocks attempts to pay off an expensive boat in New England or to feed a growing family in the Third World, overfishing has ripple effects throughout the economy.

Fresh Water

Much as ocean resources and fisheries can affect human traditions, employment, and life-styles, the fresh water that people use for drinking, washing, transportation, and to grow food can affect human health and survival.

In many cases, the problem is not a lack of water, but a lack of water free of disease or contamination. According to the U.N. Environment Programme (UNEP), dirty water or the lack of sanitation causes four fifths of the diseases in developing countries—leading to 25,000 deaths a day. At least 20 percent of the people living in cities in these nations and 75 of those in rural areas lack access to reasonably safe drinking water. And 75 percent do not have adequate sanitation facilities.[29] Even in the United States, reports the Global Tomorrow Coalition, 15 million people drink potentially unsafe water and another 30 million lack decent sanitation.[30] In the ocean section above the impact of toxic and hazardous waste contamination was made clear. These comments also apply to fresh water.

This is a problem of public health, for which engineering and public health education offers the greatest solutions. In the poorest parts of the world, the construction of a single well that reaches a safe underground water supply can save thousands of lives. After-the-fact medical care does not approach the effectiveness of engineering work to create sanitary sources of water for drinking and washing, which would prevent many diseases.

Public health thus suffers enormously in countries that simply do not have the financial, educational, and technological resources necessary to provide safe water. The connection between clean water and public health and national economic potential needs greater attention. As with the issue of species loss, our lack of knowledge impedes solution of these problems. A recent U.N. summary of the situation noted that "the lack of adequate information on water supply and sanitation at the national level is still a serious constraint to sector planning and management."[31]

Efficiency of use is a second major freshwater problem. As much as half the water in some Third World cities is lost through leaks in the delivery system, according to the United Nations.[32] Irrigation for agriculture accounts for some 73 percent of water use around the world, yet a good deal of the water never reaches crops as it is lost through evaporation or seepage from canals. Irrigation systems on average are only about 37 percent efficient.[33]

In the United States, water shortages in several regions, notably in the West and Southwest, have received considerable attention recently. Water tables around several southwestern cities and under irrigated cropland are falling as groundwater is withdrawn quicker than it can be replenished, and the Ogallala aquifer is already half depleted.[34] While this is less painful that the water problems of poorer countries, it is nevertheless important. It affects the economic potential of many regions and, therefore, the prospects for employment and well-being of their people. Damage to drawn-down aquifers can be irreversible, and water supply problems may affect future agricultural, industrial, and economic systems significantly.

The source of freshwater problems is often rapid increases in demand, inefficient irrigation in agriculture, lack of adequate conservation, poor management, and pollution. A recent survey by the Environmental Protection Agency (EPA) found that more than 10 percent of U.S. rivers, streams, and bays are significantly polluted. And more than 50 million Americans may be using groundwater contaminated by pesticides and fertilizers.[35]

Inefficiencies in the United States are encouraged by the heavy subsidization of water prices. For too long, water has been treated as a nearly free good. One third of the water supplied by the U.S. Bureau of Reclamation at low prices irrigates hay, pasture, and other forage crops of low value that are then fed to livestock.[36] In California's Central Valley, farmers pay 10 to 20 times less for their water than city dwellers do; as agriculture in that state accounts for 83 percent of the water use, this encouragement to misuse water has a dramatic impact on all residents.[37] Like the subsidies provided to those who cut down tropical forests, these pricing policies are just an incentive to waste the earth's resources.

In water-poor regions, water supplies can even be a cause for conflict. Of the 200 largest river systems worldwide, 150 are shared by two nations, and more than 50 by three to ten nations.[38] Those major rivers support 40 percent of the world's population. It would not be surprising if the next war in the Middle East—where Israel, Jordan, Lebanon, and Syria compete for access to the Jordan River—is over water, not oil.

The availability of fresh water epitomizes the cycle of life that we are part of and depend on, as the World Resources Institute explains:

> With the Sun's warming of Earth, water evaporates from the surface of land and sea into the atmosphere. The moisture is then transported great distances before it falls back to earth as precipitation. Once back in the terrestrial part of the cycle, it either flows off the surface or permeates the ground. The runoff helps to replenish rivers and lakes. The water in the ground is taken up and transpired by plants, evaporated from the soil surface or percolates downward to the water table where it is stored in groundwater aquifers. Rivers carry water to lakes and seas, and the process renews itself.[39]

As we overdraw groundwater supplies and aquifers, dump our wastes, and throw away good water through inefficient pricing and management, we imperil that natural renewal process—and ourselves.

Freshwater should be managed, to the maximum extent possible, as a finite resource on a sustainable basis. Such management requires integrated and balanced planning for water availability, population, food production, land use, economic development, conservation of natural resources, and pollution control. Use of available and appropriate conservation measures is an important component of this principle.

Agricultural Degradation and Disruption

In 1984, world agriculture entered a new age—one of failing to keep up with the world's growing numbers. As the Worldwatch Institute has noted: "Between 1950 and 1984, farmers raised world grain output 2.6-fold, an increase that dwarfed the efforts of all previous generations combined....Unfortunately, from 1984 to 1990 food output growth dropped to 1 percent per year, scarcely half that of population. This disturbing trend signals a new era, one in which an acceptable balance between food and people cannot be achieved with a business-as-usual approach."[40]

By 1989, drought-damaged harvests in key producing countries had reduced world grain reserves to one of their lowest levels in decades. When the 1989 crop failed to replenish food stocks despite better weather and high food prices that encouraged production, it became clearer that growth in world food output would have difficulties in keeping up with that of population.[41] Indeed, grain production per person peaked in Latin America in 1981, and by 1990 had dropped 16 percent; in Africa, the peak year was 1967, and 23 years later the per capita output was 28 percent lower.[42]

Increases in food output worldwide today are being slowed by environmental degradation, a diminishing response to the use of additional chemical fertilizer, and a reduction in available cropland and irrigation water. Approximately one third of the world's cropland suffers from erosion. Deforestation accentuates the problem, as water that forests no longer absorb adds to the amount of runoff from rainfall and washes away loose topsoil that, in many cases, is no longer well anchored by grasses and plants. Air pollution and acid rain damage crops as well, and some crops suffer from increased ultraviolet radiation due to depletion of the stratospheric ozone layer. Poorly managed irrigation systems and overirrigation cause waterlogging and expose crops to too much salt, lowering crop productivity.

On the erosion problem, at least, the United States has taken some important steps. The government's Conservation Reserve Program has encouraged farmers since 1985 to switch roughly one tenth of the nation's most erodible cropland over to grass or trees. The result, as noted in Chapter 4, has been more than a one-third reduction of soil loss from erosion over five years, with another drop of a third expected by 1995. (See Table 5-2.)

Balanced forms of agriculture exist, and have been practiced by indigenous peoples for centuries. Several plants can be grown side by side in order to produce a naturally balanced system, where the by-products of one plant act as nutrients for the others, and the pests dangerous to one crop are kept in check by the organisms brought to the area because of the presence of the other crops. Crops are shifted from one year to another, and fields are left fallow at cyclical periods so that they can recollect nutrients. Under this system, few chemicals are needed, because the plants themselves act as fertilizers and the ecology of the system slows the spread of pests.

TABLE 5.2: *United States: Progress in Reducing Soil Erosion from Croplands, 1986-90, With Projections to 1995*

Soil Loss	Million Tons
Excessive soil loss in 1985	1,600
Reduction, 1986-90	-600
Projected reduction 1991-95	-450
Remaining excessive soil erosion	550

SOURCE: Worldwide Institute, based on U.S. Department of Agriculture, Economic Research Services, Agricultural Resources, Croplands, Water, and Conservation Situations and Outlook Report, September, 1990.

In many countries, intensive use of chemical fertilizers and pesticides has raised productivity but failed to restore natural nutrients to the soil. In the United States, pesticides are used on 95 percent of the land planted in corn, soybeans, and cotton.[43] Sustainability of the soil drops under such chemical dependency, and the long-term potential of agriculture is distorted. Yet here, too, alternatives exist. Integrated pest management (IPM) uses the natural predators of pests, changes in planting patterns, genetic modifications of crops, and the selective use of chemicals to reduce the use of pesticides while maintaining productivity. A U.S. Department of Agriculture (USDA) study of 15 states found that farmers using IPM earned $579 million more growing nine crops than they would have otherwise.[44]

The USDA already has a highly successful program called LISA—Low-Input Sustainable Agriculture, an approach endorsed in late 1989 by the prestigious National Academy of Sciences.[45] The department has documented successes of this type of balanced, sustainable agriculture around the United States and helped the development of new sustainable farms. Unfortunately, since only some 5 percent of U.S. farmers practice alternative or sustainable agriculture, LISA has barely scratched the surface.[46]

U.S. Government farm programs and export subsidies still result in the production of some crops well beyond the level of current domestic demand. In 1989, as noted in Chapter 4, these subsidies cost the U.S. government $13.9 billion.[47] The greatest danger, however, comes from the effect on international markets after the surplus crops are dumped—the most politically attractive method for disposing of the excess supply.

American food surpluses flood international markets, depressing the prices of those commodities through a supply schedule that fluctuates suddenly. Many developing countries' economies are based on those products, and their earnings drop significantly. Third World farmers then have no incentives to improve domestic food production, because they earn too little from the crops. World prices become unstable because of the sudden, politically motivated increases in supply of specific foods.

The United States provides development assistance to developing countries through government-to-government loans and though multilateral organizations like the World Bank. But the cost to developing countries of depressed and fluctuating food prices that reduce their income from agricultural crops can be greater than the benefits of development assistance. And depressed commodity prices have destabilizing effects that cause lasting damage to economic potential as well, such as the disincentive to produce food when it earns too low a price in the market.

EC farm programs and export subsidies in Europe have been a major source of conflict in worldwide trade negotiations under the General Agreement on Tariffs and Trade (GATT). The Uruguay Round, the most recent set of negotiations on GATT, has been slowed by conflict over agricultural policies. Domestic politics in European countries have made it difficult for governments to reduce expenditures on farm programs and export subsidies, while some other governments demand that they be changed.

Overall, there is no question that modern agriculture and the Green Revolution have raised food outputs in the Third World. But the effects on the land and on society are not necessarily positive, as Bunker Roy of the Social Work Research Centre in India describes. (See Box 5-2.) As the world continues its push to feed 5.4 billion people—soon to be 6 billion and then at least 7 and 8 billion—it must deal with both the degradation of agricultural resources and the disruption of social systems.

Hazardous Substances

The term "hazardous" waste is in some ways an arbitrary legal one. Many "non-hazardous" waste streams generated by the public, utilities, and industry are toxic or otherwise harmful, but are not defined or regulated as such. So we do not know the full extent of the hazardous waste problem. But what we do know is frightening enough. The United States produces more than 260 million metric tons of regulated hazardous waste each year—more than a ton for every person in the country. In several states, the annual per capita figure is more than two tons.[48]

These materials can cause a wide range of harmful effects on human health, as well as permanent damage to ecosystems. Moreover, manufactured toxic chemicals can accumulate in the environment over time, and they can travel in the atmosphere, in ocean currents, and through groundwater.

The Environmental Protection Agency recently released results of the first U.S. inventory of toxic substances discharged into the environment, a study that was made possible by the community right-to-know provisions of SARA, an act addressing the management of toxic substances. Its report

Box 5-2: *The Impact of Agricultural Mechanization in India*
When bullocks and wooden ploughs in India suffice for over 80 million small farmers owning less than five acres of land, it is ludicrous to suggest that tractors should replace them. This would destroy the traditional cycle in which farmyard manure is returned to fertilize the fields, would eliminate the milk and butter provided by the animals, and remove their labour from other agricultural work. The tractor not only requires expensive fuel, it also produces a system of farming that demands chemical fertilizers to replace the no-longer produced farmyard manure, the purchase of milk and butter, and the use of other machines: electric or diesel pumps, for example, to replace the bullock-power which previously pumped irrigation water.

Advising small farmers to go in for tractors forces them to enter a life-style they can neither afford nor control. Nor can they stay small: more land is required to justify the tractor; poor farmers unable to afford tractors are squeezed from their farms; there is a reduction in the need for both permanent and casual rural labour. In short, the introduction of tractors forces a dramatic and far-reaching alteration in the life-style of the farmers who buy them and also undermines the sustainability of the community into which they are introduced.

SOURCE: Excerpted from Bunker Roy, "Global Connections 2," Social Work Research Centre, Tilonia, Rajasthan, India, undated.

---

found that the nation's industries released 10.5 billion pounds of toxic substances into the air, land, and water in 1987, including 550 million pounds that was dumped into water. More than half the compounds discharged into streams, lakes, and rivers were not covered by EPA regulations. The agency found that 10.5 million pounds of carcinogenic arsenic compounds were released into U.S. waters in 1987.[49]

Today, hazardous wastes have become an export item. Poor countries desperate for foreign currency have been known to agree to be dumping grounds. Greenpeace estimates that at the very minimum 10 million tons of toxic waste were exported between 1986 and 1990, with more than half of it ending up in Eastern Europe or a developing country.[50] And the United Nations Environment Programme says that about 800,000 tons are shipped within Europe annually.[51] An international agreement that is still awaiting ratification by enough countries to go into effect—the Basel Convention—would restrict transboundary movement of waste without notification by the exporter, consent of the importer, documents of verification, and proper disposal in the recipient country.

In the case of hazardous waste, the problems are clear. But in many other situations it is difficult to determine whether a product is beneficial or harmful. Chemical products, for example, offer benefits ranging from increased life expectancy to expanded economic opportunity. Yet the environmental and health impact of chemicals and their by-products are severe. Measuring the benefits and costs of chemicals is a subjective process, and depends on good data, which are often lacking.

Chemicals represent about 10 percent of the total value of world trade. Some 70,000-80,000 chemicals are now in world markets. Every year, 1,000-2,000 new chemicals join them, many without adequate prior testing or evaluation of effects. A U.S. National Research Council sample found that complete health hazard evaluations were available for only 10 percent of pesticides and 18 percent of drugs in the United States. Toxicity data existed for only about 20 percent of the chemicals used in commercial products and processes inventoried under the Toxic Substances Control Act.[52]

The capacity to test drugs and chemical substances in the United States does not approach the pace of their production, and in most other countries the situation is even more hopeless. People are exposed to a copious quantity of chemicals that they do not understand.

John Todd of Ocean Arks International, who works on ecologically sound methods for treating waste, notes that:

> The waste-treatment industry is getting scary. To meet regulations on some chemicals, we use others that aren't regulated. We use chlorine to meet ammonia standards, and in the process make chloroform and chloramine, which don't have standards. To get rid of phosphate, we precipitate it out with aluminum. Aluminum is toxic in all parts of the environment, but we haul it out of sewage-treatment plants and dump it onto the land by the ton. We use high concentrations of copper salts, which are not natural in ecosystems, to get rid of algae, which are. Every time the restrictions on one pollutant get stronger, the chemicals to remove it get stronger.[53]

This proliferation of unregulated chemicals to control regulated ones illustrates well the difficulty of trying to fiddle with just one part of a problem: The solution can bring its own headaches. A better goal is to reduce the use of chemicals overall. Such a policy of toxics use reduction, as demonstrated by programs in Massachusetts and Oregon, would encourage industry to undertake changes in production processes, final products, or raw material use so as to reduce, avoid, or eliminate the use of toxics per unit of product.[54]

Whether a chemical is accepted or banned by the U.S. government, it can often still be exported. As with hazardous wastes, developing countries that have no institutions to monitor the products entering them serve as the markets. In some cases, this may be reasonable, as those countries have different needs and problems than the United States does, and may benefit more from a chemical than Americans would. In other cases, the citizens and environment of developing countries suffer intensely from the use and misuse of chemicals. A 1985 survey in one state in Brazil found that 6 out of 10 farmers who used pesticides had suffered from acute poisonings.[55]

Dangerous chemicals used in foreign countries affect Americans as well. Pesticides applied to crops in one country are imported into another, and pesticides sold by Americans on international markets can find their way back into the United States in food products, in what has become known as the "circle of poison." In 1987, for example, the USDA stopped 30 million pounds of Australian beef from being marketed because of illegally high residues of DDT and dieldrin.[56]

As noted in the preceding section, moving toward sustainable agriculture would reduce the use of these threatening substances. It would be a sort of "source reduction" strategy for pesticides that could parallel industry's efforts to cut toxic emissions. In both cases, the goal is reduced stress on the environment and fewer risks to human health.

Natural Disasters

Although natural disasters often cannot be avoided, their severity depends on how prone people and systems are to the effects. In the late 1980s, earthquakes struck northern California and Armenia in the Soviet Union. Yet a comparison of the damages showed an enormous inequality between the two regions: In California, damage was largely contained by the quality and design of the architecture, which often held up under the pressure and therefore preserved lives. In Armenia, this was not the case, and buildings collapsed inward.

Similarly, people who have put environmental systems under stress become more prone to disaster, mainly drought and floods.[57] As Lloyd Timberlake of the International Institute for Environment and Development has noted:

> To the average newspaper reader, the causes of Africa's troubles in 1983-1985 seemed tediously obvious: there was a shortage of rain, leading to poor harvests, leading to famine. The rains would eventually come again, and the troubles would be over.
>
> But rainfall was not the whole explanation of Africa's hunger....A drought is a lack of water, but not necessarily a disaster. Whether or not a drought becomes a disaster depends on how people have been managing their land before the drought.[58]

Plants and trees store water and nutrients, letting them out into the soil in times of low rain, and storing them in times of high rain. Increasing pressures due to population growth and poverty in Sahelian Africa led to the cutting of trees for fuelwood to heat and light houses, and prevented trees from playing this moderating role. When rain became short, no relief came from the natural systems of water and nutrient conservation, and famine was accentuated. Timberlake continues:

> In the case of floods, people make the disaster itself more likely, by clearing vegetation and compacting the soil so that the land loses its 'sponge effect'—its ability to soak up large quantities of rainwater and release it slowly. Given the rapid rates of world deforestation, it is not surprising that floods are the fastest growing natural disaster: 5.2 million people affected per year in the 1960s, 15.4 million per year in the 1970s.
>
> Like floods, droughts are largely caused and exacerbated by human action. Though apparent opposites, droughts and floods are closely related. Droughts are indeed partly caused by too little rain, and floods by too much, but both are also caused by the land's inability to absorb rain.[59]

In Bangladesh, artificial changes in natural systems have similarly caused environmental stress and contributed to vulnerability. A sea-level clearing of mangrove forests for rice paddies destroyed a natural barrier to the sea. When a tidal wave hit the unprotected coast in 1970, at least 300,000 people died. Since then, Bangladesh has begun mending its sea fences with a large-scale mangrove replanting project.[60]

Politics and economic pressures can also increase proneness to natural disaster. During the drought year of 1983, and the following year, 1984, the five Sahelian countries hardest hit harvested a record 154 million tons of cotton fiber. The crop was grown largely by small farmers—the same ones who could not grow enough food to feed themselves. While exporting that cotton, the Sahel set another record in 1984 by importing 1.77 million tons of cereals. The fact that cotton could be grown but grain could not had more to do with government and international aid policies than with rainfall. In order to pay back foreign debt incurred from development projects, agriculture in the Sahelian countries had been shifted toward cotton for the foreign currency it would yield and away from food for local consumption.[61]

Modern systems have the ability to improve people's protection from the unleashed power of nature. Instead, however, they have sometimes disrupted balanced natural systems that offered shelter and safety throughout history. And patterns of international trade that leave people dependent on imports of food paid for with money earned from exports of a narrow selection of crops or natural resources leave people similarly vulnerable. It is important to distinguish between trigger events—such as hurricanes or too little rain—and the associated disasters, which can be caused by putting environmental systems under stress or leaving people vulnerable for political or economic reasons.[62] Otherwise, relief efforts will continue to supply bandages rather than preventive care.

Atmospheric Issues

German Astronaut Ulf Merbold, after looking from space at the earth, said "For the first time in my life I saw the horizon as a curved line. It was accentuated by a thin seam of dark blue light—our atmosphere. Obviously, this was not the ocean of air I had been told it was so many times in may life. I was terrified by its fragile appearance."[63]

Many others are terrified by what we are doing to that fragile layer. Air pollution, ozone depletion, global warming. The public has become so familiar with these facts of life that you can hear people on the bus talking about seemingly abstract scientific topics on the way to work. Reducing the stress that human activities are putting on this overarching environmental system is perhaps the most basic and challenging global issue of the next decade.

Pollution was the first truly international environmental issue of our times. As noted earlier, rivers and streams carry wastes across borders, pulling nations into treaties to protect their waters. Similarly, airborne pollution travels the globe, affecting all countries regardless of its country of origin. It finds its way into ocean ecosystems, plants, and animals.

Short-term exposure to high levels of air pollution and long-term exposure to low levels can cause a variety of illnesses in people, from lung disease and cancer to fetal defects and exacerbation of existing heart disease. Infants, children, the elderly, and people with respiratory problems are most at risk. According to the EPA, 150 million Americans breathe air considered unhealthy.[64]

Smog is formed when strong sunlight acts on a mixture of nitrogen oxides and volatile organic compounds. Cars are a major source of nitrogen oxides. Although improvements in emission controls and other technologies over the last two decades should have reduced smog in the United States, the effects were offset between 1977 and 1988 by the 25-percent increase in cars and 40-percent increase in trucks. Smog has increased significantly in many urban areas.[65]

Other types of damaging air pollution include sulfur dioxide, carbon monoxide, hydrocarbons, ozone, carbon dioxide, and particulates. Particulates are solid and liquid materials that remain suspended in the air. They come from aerosols, forest fires, the burning of certain fuels, and other sources. The U.S. Office of Technology Assessment estimates that current levels of particulates and sulfates in the air may cause the premature death of 50,000 Americans per year, accounting for 2 percent of annual mortality.[66]

The World Resources Institute notes that air pollution has decreased in numerous areas over the last 20 years: Concentrations of sulfur dioxide, for example, declined in 20 of the 33 cities monitored by UNEP. Sweden cut its emissions of this pollutant by two thirds between 1970 and 1985, while Germany did the same just between 1983 and 1988.[67] The U.S. record is less impressive—emissions of sulfur oxides dropped 28 percent from 1970 to 1987.[68]

At the same time, however, things have gotten decidedly worse in many developing countries, where cars, buses, and trucks continue to leave noxious clouds in their wake and regulation of industrial emissions is just beginning. Breathing the air in Bombay is equivalent to smoking 10 cigarettes a day, and the scratchy throats and stinging eyes suffered by those living in Mexico City are only the most obvious signs of the infamous polluted air there.[69] Overall, the World Health Organization estimates that 70 percent of city dwellers, mainly in the Third World, breathe air that has unhealthy levels of suspended particulates at least some of the year.[70]

The news regarding the ozone layer is not good either. The depletion of the thin shield in the upper atmosphere that protects the earth from ultraviolet radiation is proceeding faster than scientists at first thought. Data released in April 1991 indicated that the loss was proceeding twice as fast as had been expected over the United States and nations at similar latitudes. EPA Administrator William Reilly called the discovery "stunning information" and "disturbing."[71]

Every percentage point of ozone depletion translates into a 5-7 percent increase in skin cancers, according to researchers.[72] For Americans, the new data mean that over the next 50 years about 12 million people will develop skin cancer, which some 200,000 of them will die from.[73] In the southern hemisphere, over which the ozone layer has been eroding longer, the situation is such an accepted part of life that Australian children now learn not only about the 3Rs but also the 3Ss—slip on a T-shirt, slap on a hat, and slop on the sunscreen.[74]

The United States uses 20-30 percent of the two main causes of ozone depletion—two forms of chlorofluorocarbons (CFCs).[75] So it was entirely appropriate that the government played a leading role in the 1987 negotiations that led to the Montreal Protocol, which called on nations to cut CFC emissions in half in a decade. By June 1990 in London, the accord signatories agreed that industrial countries should stop all production and use of ozone-depleting substances by the year 2000, and that developing nations should do the same by 2010. The new information released in 1991 is bound to increase pressures to speed up the timetable even more.

The U.S. role in discussions on the most worrying atmospheric issue could not be more different from what happened in Montreal. Burning fossil fuels releases carbon dioxide (CO_2) into the atmosphere. Combined with carbon dioxide, CFCs, and gases from other sources, this accumulates and forms a layer of gas that traps solar radiation near the ground, warming the globe and ultimately changing the earth's climate.

The results could be disastrous: global mean temperatures could rise by 1.5-4.5 degrees Celsius by 2030, and sea level could rise 30 centimeters over 50 years.[76] Two key points must be made here. First, this oft-cited temperature increase assumes only that atmospheric concentrations of CO_2 double and then stabilize. Obviously, however, more warming will occur if those concentrations more than double. Second, no country's temperature will increase by the global mean. The actual range of increases will result in even higher temperatures in several countries. We are only just beginning to realize the many devastating effects this could have on food production, biodiversity, energy use, and the lives of millions of people residing in low-lying areas.

The United States, with just 5 percent of the world's population, is the leading producer of the greenhouse gases responsible for global warming. In 1987, it accounted for some 18 percent of the world total.[77] Barring any actions to change this situation, according to the Office of Technology Assessment, U.S. CO_2 emissions are likely to rise 50 percent over the next 25 years.[78] Yet the United States remains alone among industrial nations in not accepting the seriousness of this problem. Some 23 countries have already established goals for freezing or reducing CO_2 emissions, and many have called for an across-the-board cut of 20 percent in the climate convention now being negotiated.[79]

Scientific uncertainty about global warming has been misused as an excuse for political inaction and unpreparedness for possible future needs. Although we cannot be sure about the timing and extent of global warming, most researchers and many governments endorse the position of the Intergovernmental Panel on Climate Change (IPCC). In 1990 the IPCC, set up by UNEP and the World Meteorological Organization, reported that to avoid further warming CO_2 emissions must be cut by 60 percent right away.[80] Germany, a country twice as efficient in its use of energy as the United States, is already moving toward a plan that would result in an 80-percent reduction over the next 60 years.[81]

The IPCC endorsed the immediate adoption of numerous short-term options to reduce energy use that are "beneficial for reasons other than climate change and justifiable in their own right"—a position characterized as a "no regrets" response to global warming. Examples include vastly improved energy efficiency, afforestation programs, and an accelerated phaseout of ozone-depleting chemicals, which also have significant potential to contribute to global warming.[82] Regrettably, the U.S. government does not agree with the IPCC position.

The rapid development by industry of CFC substitutes that followed the signing of the Montreal Accord is indicative of what can be accomplished when the global community decides to force change. Although the modifications needed to mitigate global warming affect society more broadly, the technologies to lower energy use already exist and policy directions to encourage it are well known. Many industrial nations are moving rapidly to cut CO_2 emissions. The refusal of the U.S. government to join in can only be called unconscionable.

CHAPTER 6

Social & Cultural Systems

Social and cultural systems are the means people use to take care of their needs for health care, education, housing, religion, art, communication, and myriad other interactions.

Much as ecosystems function through interconnected cycles, social and cultural systems rely on and support each other. And just as ecological problems often stem from a breach in those cycles, so do social and cultural problems often arise from broken links.

The poor availability of education or health care, for example, is tragic for its own sake, but it also has implications far greater. It reduces the potential of all the human endeavors that rely on healthy, educated people, and contributes to a circle of decline that further reduces the quality of health and education, onward in a downward cycle.

Similarly, improvements in the basic human condition project themselves far beyond concerns such as health and education alone. They unleash human creativity and innovation in ways that can cause improvements in any area, including medicine, science, and public affairs. The relationships are almost ecological in their interactions: Increased hopefulness moves along the paths created by greater education, which opens opportunities for constructive employment, which brings in money and insurance for health care and improved diets and housing, reducing the pressures that lead to environmental degradation and societal disintegration, and establishing an upward cycle that builds toward security for individuals and society alike.

But current thinking holds that social policies are wealth-consuming and that economic policies are wealth-creating. Under this concept, economic expenditures often receive priority, in the name of long-term national well-being, over social expenditures. Yet the division between the two is a false one, and wherever social goals do not receive their due, other priorities suffer as a result.

The roots of many problems lie in a failure to recognize the crucial role of social and cultural systems in supporting all activities. From the development assistance that countries provide to the way they treat their own citizens at home and the underlying patterns of the economy, nations shape the future when they make decisions about social priorities.

Health Care

Not surprisingly, good health is the foundation of a healthy economy. When people cannot work up to their potential, when they must struggle to fight off illness from the water they drink, the air they breathe, and the work they do, they lose the ability to contribute fully to society as well as to their own well-being. In much of the world, this struggle goes on day after draining day.

In many parts of the developing world, the health care system is not ony under stress, it barely exists. In the mid-1980s, there was one doctor for every 60,000 residents in Ethiopia, for every 6,290 people in Thailand, and for every 1,240 Mexicans. In the United States, by comparison, the ratio was one for every 470 people.[1] Those living in rural areas are even less likely to receive care; in the 62 countries classified by the U.N. Development Programme (UNDP) as having low human development, 40 percent of the rural population has access to health care services, compared with 85 percent of the urban population.[2]

Women are particularly vulnerable to a lack of access to health care, the result of a long-standing undervaluation of their contributions and their roles in economic life. Some 1 million women die each year of reproductive causes (pregnancy, abortion, sexually transmitted diseases, reproductive cancers), and another 100 million suffer disabling illnesses. In most developing countries, 20-45 percent of all deaths among women in their childbearing years are connected to a pregnancy. According to Worldwatch Institute, African women face a 1-in-21 chance of dying from such causes during their lifetime. In North America, in contrast, the risk is 1 in 6,366.[3]

Children are similarly susceptible to disease because of poor access to health care or a complete lack of this basic societal support system. According to UNICEF, some 25,000 children under the age of five die each *day* of preventable diseases, such as diarrhea, whooping cough, and measles.[4] Forty-one percent of children this age in UNDP's category of least developed countries are classified as underweight and suffering from malnutrition. With an overall infant mortality rate of 99 per 1,000 live births in these nations, a good number of infants do not even get a chance to face these difficult conditions.[5]

Poor health, and thus a weakened ability to contribute to building a sustainable society, is not confined to the developing world. As the wealthiest nation on earth, the United States has more ability than any other country to secure good health for its people. Yet the commitment to improving health in the United States is lacking in many ways.

Safety in the workplace, infant mortality, availability of health insurance, exposure to toxic chemicals in the environment and in buildings, indoor air pollution, automobile safety, and other health concerns are all currently below the quality of health that Americans could achieve.

The Institute for Southern Studies in Durham, North Carolina, recently reviewed a wide range of environmental health conditions in all 50 states. Researchers Bob Hall and Mary Lee Kerr ranked states on such measures as cancer deaths and rates per capita, homes without adequate plumbing, infant mortality rates, population in medically underserved areas, and occupational deaths. Their results revealed a troubling picture:

> The 23 indicators we chose to measure a state's community and workplace health emphasize [the] relationship between public policy, pollution, living conditions, and human health. The results are distressing—the nation's community and workplace health is ailing and no state can claim a clean record. Not a single state ranked in the top half in all indicators....One in seven of the nation's non-elderly population—33 million people—had no medical insurance coverage in 1988, either through a private plan or Medicaid. The proportion jumps to one in five for African American children and almost two in five (38 percent) for Latino children, trends that translate into fewer doctor visits. Regardless of race, families with

incomes under $25,000 are only one third as likely to have insurance coverage for their children as families earning over $40,000.[6]

According to Hall and Kerr, in 1988 the Bureau of Labor Statistics calculated that about 11,000 workers were injured seriously enough each day to lose work time or be put on restricted work duty. The rate of lost-time injuries has increased by 39 percent since 1974. They conclude that "on a national scale, the health data suggest that workers and the poor are considered disposable."[7]

The death rate in the United States for infants under the age of one is 9.6 per 1,000 live births. Every day more than 100 American babies under one year old die—a total of 39,500 infants in 1989.[8] In this respect, the United States fares far worse than other industrial countries. The infant mortality rate in Japan is 4 per 1,000 live births, while in Sweden and Finland it is 6 and in Canada, our neighbor, it is 7.[9]

Cancer rates, and rates of other diseases, vary highly among the different states. Those with old industrial areas and with poorer medical care have higher rates. In local regions near chemical industries, cancer rates run very high. These findings show that regulation of industries, provision of health care, and public health actions in general have a positive effect in protecting people.

As in other matters discussed in this volume, for the United States to exert leadership worldwide in matters of health it will have to improve its record at home. No nation can afford to treat its workers and its impoverished population as disposable. Or to lose the potential contributions of countless women and children who die needlessly. A recognition of the role that healthy people from all areas of a country and all occupations play in national well-being is an important part of national policy.

The Connection Between Health and Education

The connection between education and public health deserves a greater focus than it has received. It highlights the ties among social priorities, and their mutually beneficial nature.

A farmer who cannot read the label on a pesticide can is in danger of being poisoned. A worker who cannot fathom the meaning of instructions on a new piece of machinery can be injured or even killed. A mother who cannot understand the description of how to mix oral rehydration salts or give other types of medicine puts her child's health at risk. And if she cannot read the instructions on a packet of birth control pills, or understand the contraindications for taking them, she is at risk herself.

With illiteracy rates of 60-80 percent in a good number of developing countries, this connection between education and health is a very direct one. Education thus has the potential to improve public health. Today's medicine can cure almost all types of diseases, including cancer. But medical care depends on early diagnosis; even in the United States, many people resist acknowledging potential problems and do not know how to detect them early.

Schools in the United States often do not teach the fundamentals of how the human immune system operates. Yet it is the mechanism through which all ailments from the common cold to the most life-threatening disorders are fought. The schools lose the opportunity to educate young people on diseases that their friends and relatives, and even themselves, will be coping with in the future.

Historically, the largest improvement in public health of all time came with education about the importance of sanitation and clean water for drinking and washing. Today, education in the United States is needed about the diseases that face us in larger and larger numbers: cancer, heart disease, and AIDS. It is also required regarding automobile accidents and seat belts, smoking, the role of stress in reducing immunological potential, drugs, and chemicals and pesticides in food.

Major changes have taken place in this country in education about the dangers of car accidents and cigarettes. Seat belts are now a mandatory part of all cars, and their use is required in many states as well. Americans have a greater tendency to wear their seat belts than people abroad. Similarly, smoking has declined enormously in recent years in the United States, and is far less common than in almost all other countries.

People are becoming more aware of the significance of stress all the time, and are trying to reduce it where they can in their lives, although that is not always possible. But abuse of drugs in on the rise, and encompasses both the use of illegal drugs and the misuse of prescription ones. And the addition of chemical additives to foods and the application of chemical fertilizers and pesticides epitomizes the use of a continual stream of new chemicals, many of which have the potential to cause the diseases from which Americans suffer the most.

Public health is often affected by decisions over which chemicals to manufacture and use, and over how products are regulated. But people have little or no information about the procedures used by the Food and Drug Administration (FDA) as it regulates chemicals, foods, and drugs. Often, even the FDA lacks complete information, and scientific uncertainty is present in most cases over the effects of new products. As noted in Chapter 5, some 1,000-2,000 new chemicals enter the market each year, many without adequate testing.[10]

Decisions over these vital issues have been delegated to the Public Health Service, and people outside that system do not participate significantly in its priorities or procedures. Education over the government's actions in chemical and food standards, and their links to public health, is difficult to come across and certainly not a part of the daily reading of the American public. Yet the links demand that a coherent and comprehensive strategy be adopted in the face of potential influence by special interest groups and the hindrances of bureaucracy over the selection of standards.

Similarly, environmental health in general is linked to public health, yet no comprehensive environmental plan exists in the government, and environmental decisions are divided in a fragmented manner among various government agencies.

Physicians cannot possibly meet all the health needs of a population. Those must be met largely through avoiding health problems by maintaining a clean, healthful environment, teaching people how to protect their own health and when to seek medical advice, and providing good sanitation and clean food and water.

Poverty in the United States

Inequality turns any country's most fundamental asset—the potential of its people—into a liability. Poverty acts as a stress on society. Just as personal stress reduces the ability of the human body to fight off diseases and live at full health, stress in a society undermines the support systems that maintain its health.

It damages social cohesion, breaking neighborhoods and cities into fragments that contradict the interconnected patterns of healthy, mutually supportive systems. It directs activities that could have been constructive into actions that are wasted, or even destructive, such as crime or drug and alcohol abuse. Social stress challenges the ability of communities to provide enough of such basic services as medical care and housing to their people.

Poverty may begin with the failure of economic systems to meet the needs of all members of society, but in the United States it goes beyond that and becomes all at once problems of education, public health, racism, discrimination against women, housing, employment, crime, and personal hopelessness.

Although by the end of the 1980s, the number of millionaires in the United States reached 1.5 million, the bottom 40 percent of the population has experienced a decided decline in their income after taxes over the last 15 years.[11] The Brookings Institution reports that the share of the national income going to the wealthiest 1 percent rose from 8.1 percent in 1981 to 14.7 percent just five years later.[12]

Meanwhile, about 33 million Americans—13.5 percent of the population—live below the official poverty line set by the Bureau of the Census each year, which in 1990 stood at $13,359 for a family of four.[13] And nearly two in five of them have incomes less than half the poverty threshold, an increase in this group of "hyper-poor" of nearly 45 percent since 1979.[14] Recent debate over how the poverty line is calculated indicates that a more accurate measure would find closer to 18-20 percent of Americans living in poverty.[15]

Women and children, especially if they belong to a minority group, are particularly hard hit. Forty percent of those living below the poverty line are under the age of 18, and just over half of the poor are single-parent families headed by women.[16] In an analysis of data from industrial nations in the early 1980s, the United States had the highest poverty rate for children; Australia, with a per capita income 20 percent below that of the United States, had a lower proportion of children living in poverty.[17]

As the lines of people waiting for a soup wagon's daily visit lengthen, and as the newspapers carry stories of bulging shelters for the homeless, often filled with families who find themselves falling out of the middle class, it becomes clearer that this is a kinder, gentler nation only for some. And that the country needs to revise its ill-fated war on poverty.

Programs that offer training and create jobs where they are most needed lead ultimately to the security that all people crave and that unleashes their creativity and imagination. A lack of jobs, education, and health, on the other hand, quashes creativity and motivation among people who have so few opportunities that they have resigned themselves to unemployment and isolation from the rest of society.

To prevent even more people from being caught in the poverty trap, training is also urgently needed for those whose incomes depend on environmental systems that are under growing stress. As discussed in Chapter 5, fishers in the New England states are being forced out of business by depleted fisheries. People dependent on the logging industry in the Northwest similarly face considerable changes in employment—not only because of environmental concerns about old-growth forests but also because of mechanization and an increase in raw-log exports.[18] And farmers trying to get off the pesticide treadmill and practice sustainable agriculture are fighting an uphill battle without support programs.

Recent foreign policy triumphs and rapid changes in the global political scene have produce a sense of euphoria about the position of the United States in the world. But the reality of the domestic problems facing this country will set in soon for more people, as it has already for the one fifth of Americans who are at the bottom of the ladder. As one of the wealthiest countries in the world, the lack of dedication to solutions for the problem of poverty on the part of the United States represents a fundamental failure to recognize the importance of balanced, mutually supportive development.

Undervalued and Disadvantaged Groups

In an interconnected system, all parts play a vital role and offer strengths that the others rely on. But societies today tend to undervalue the contributions made by several key groups—women, minorities, and indigenous peoples. They are underpaid in the workplace, receive fewer opportunities, and have less of a say in political and other types of decision making.

The roles that women play are particularly important not only to the health of their own families and the economy of their nations, but also to the health of the environment that supports us all. And they are disproportionately affected by stresses placed on that environment. Women in Africa provide 60-90 percent of the subsistence agricultural labor force, and produce 70 percent of the staple food.[19] They often do not have the luxury of doing this in a way that preserves the natural resource base for future generations.

Although women's work accounts for two thirds of the hours of productive activity around the world, they earn just 10 percent of the total income.[20] In developing countries, women often spend most of their waking hours working. Yet the access they have to health care, legal rights, schooling, and jobs is severely curtailed in many parts of the world. The stress this in turn places on children is tragic in its own right, and particularly worrying as it leaves them ill prepared to deal with the problems of the next generation—problems that are only going to get worse if we fail to address them now.

Improving the health, education, and status of women is thus a key to building a sustainable future. In the United States, achieving this is closely tied to solving the problem of poverty. As noted earlier, 52 percent of poor Americans live in single-parent families headed by women, a figure that has more than doubled over 30 years.[21] Providing women with more equal access to all society's resources would acknowledge their roles as agents of change. Rather than seeing them as part of the problem, it would begin to tap their potential to contribute to solutions.

These special roles were recognized at a recent symposium on "Women and Children First," held in preparation for the 1992 Earth Summit. Participants called for "a recognition of the unique health problems faced by women and children due to unclean water, unsanitary household or community environments, depleted energy sources, land degradation, environmental pollutants, toxic waste disposal, toxic chemical use and other unsound practices that are preventable." They further called for "a recognition of the actual contributions of women in environmental protection and development, especially their role as protectors and managers of fragile natural resources."[22]

The diversity of cultures in the United States has been one of its greatest strengths historically, but the members of those cultures are often blocked from contributing their approaches and knowledge to solutions to today's problems. What could have been treated as a resource has been treated, instead, as a liability, and members of minority groups—African Americans and Hispanics, in particular—have too often found themselves impoverished, living in slums, and dependent on welfare programs of the government. The ratio of black to white unemployment is larger now than it was in 1960. The overall unemployment rate for African Americans in 1987 reached nearly 13 percent, and in many inner-city areas it stood much higher.[23]

All the wastage of human potential described earlier that is inherent in poorer education, lower paying jobs, and less access to adequate health care is magnified in the case of minorities in the United States. And the frustration it breeds translates into stresses on individual health, family life, neighborhood and personal safety, and cultural cohesion. To unleash the full creativity and innovation potential of all its members, society must break the downward cycle that has trapped those most discriminated against.

Indigenous peoples the world over developed cultures and life-styles that stayed impressively in harmony with natural systems for millennia, and that passed on knowledge of natural methods of healing, growing crops, and using the environment sustainably. Some 250 million people around the world can still be considered indigenous—descendants of those who first inhabited land since colonized by foreigners.[24] Their ancestors knew how to balance demands on the earth with their own demands for food and sustenance, and many of them still follow those practices.

Whether it is a few thousand San Bushmen who still move around the Kalahari Desert in Africa hunting animals, or Pacific Islanders who search for octopus and turtles in canoes carved from breadfruit trees, or the Hanunoo of the Philippines who have identified 1,600 plant species in the forest that they can use, indigenous people know that the earth is the source of life, not merely an economic resource.[25]

Yet the role of these groups as the original practitioners of sustainable development is rarely acknowledged. And as their societies come under increasing attack, their valuable knowledge is being lost. Their connection to the earth is being severed. Hayden Burgess of the World Council of Indigenous Peoples describes the result: "Next to shooting indigenous peoples, the surest way to kill us is to separate us from our part of the Earth. Once separated, we will either perish in body or our minds and spirits will be altered so that we end up mimicking foreign ways, adopt foreign languages, accept foreign thoughts....Over time, we lose our identity and...eventually die or are crippled as we are stuffed under the name of 'assimilation' into another society."[26]

The plight of such groups as the Yanomami and the Kayapo in Brazil and the Penan and Dayak tribes in Malaysia are at last receiving some much needed international attention. And perhaps this will help raise awareness about the many other groups whose cultures and livelihoods are being overrun by the rush to pay off debts and raise the gross national product. Whether the attention comes too late to save their cultures—and their knowledge of how best to use the many resources of their rain forests—remains to be seen.

The issues should not be thought of as affecting only "primitive" groups in faraway lands. Some 1.5 million people in the United States are considered to be members of indigenous groups, including the Aleut, native Alaskan, and Inuit in Alaska. They are the most disadvantaged people, suffer the worst health, receive the least education, and among the poorest groups in society, according to *The Gaia Atlas of First Peoples.* The atlas notes that mining projects are a source of conflict among the Hopi and Navajo, that the Shoshone are subjected to weapons testing on their lands and to serious human rights violations, and that environmental degradation and pollution are placing stress on the Iroquois and Lakota.[27]

The valuable contributions that these Native Americans have made and could continue to make to the sustainable development of the nation is a resource we can no longer afford to waste.

International Lending

Today, organizations dedicated to economic change have a major impact on social and cultural systems, and in many cases are adding to the stress on those systems. International lending institutions, such as the World Bank and the other multilateral development bodies make decisions that affect the education, health, employment, family size, and housing of people. Even personal security is involved, as those variables are so important as to be threatening when they are insufficient.

Social policy is part of the mandate carried by national governments and decided through domestic politics. But in the countries where development banks have more money to spend than national governments do, and where development is desperately needed, social policy can take a backseat to development projects. And it can be decided in part by the actions of international institutions.

This is a historic situation, and it is a sign that national sovereignty is changing. International politics wields more power than ever before in changing the ways of life of people who have never traveled outside their countries or sought out a place in the global economy. Unfortunately, these people usually have no voice in international politics or economic policymaking. The development projects that can affect them most directly are conceived in their own capitals or those of some other nation, decided on by a board of directors made up of people from other countries who have never visited the project site or in most cases encountered the local culture, and implemented largely by foreign consultants and contractors using foreign technology and plans.

Development projects change local systems of education, public health, transportation, food production, energy consumption, housing, and employment. But many are designed and evaluated almost completely by people with just one background: economics. They use economic models that predict the success and effects of each project. They have often, however, been surprised by the results. Projects that made sense from an economic point of view have caused ecological damage severe enough to have a boomerang effect that harmed the development project itself. Or they dislocated enough people to undermine the human resource base that the project depended on, thereby reducing its success and also calling into question its purpose. The number of noneconomic changes introduced by development projects that made sense from a purely economic point of view is almost without limit.

One of the most famous and contested development projects is an irrigation and hydroelectric power scheme on the Narmada River in India funded by the World Bank. Costing tens of billions of dollars, the project is intended to go on for 50 years and include more than 3,000 dams. It will eventually force more than 2 million people to relocate from their homes.[28]

The 67,000 people to be forcibly relocated during the first phase—the Sardar Sarovar Project—are supposed to receive land equivalent to what they lose, but there is not enough land in the area to meet that goal. They will probably move onto the hills surrounding the lake that the dam will create, where deforestation and erosion are already serious. With the increased population pressure, erosion and siltation of the reservoir may diminish significantly the chances the project has for economic success.[29]

The Sardar Sarovar project would flood 875,000 acres of forest, but the economic calculations that the World Bank carried out to justify the endeavor failed to account for the many services provided by the basin's natural resources. These include soil conservation, climatic regulation, water conservation and replenishment, and wildlife habitat. Cost-benefit calculations also failed to consider such expenses as controlling the waterborne disease of schistosomiasis, and those costs to human well-being and productivity can be spectacular.[30]

Intense protests at both the local and international level have gone on for years, and signs are now beginning to appear that they have had an effect. It is possible that the combined pressure of many different groups will persuade the Bank to significantly change its actions in the Narmada Valley and to include ecological, public health, and relocation concerns in its planning of projects there.

Another famous and catastrophic scheme is the Polonoroeste Project, a loan of nearly a half billion dollars from the World Bank to fund Brazilian Highway 364 through the Amazonian rain forest. More than 200,000 migrants traveled over it in 1985 alone, most of them lured by the promise of free land and the possibility of making a livelihood. Even those fortunate enough to receive land, however, found that rain forest soils once cleared of their vegetation are unsuitable for the agriculture they knew how to practice.[31] Contrary to government and Bank expectations, only 9 percent of the state of Rondonia turned out to be capable of sustaining agriculture.[32]

Those drawn to the area cleared the tropical forest, planted it for a year or two, and watched their crops fail. To survive, they were forced to work at subsistence wages for cattle ranchers or speculators or to move on to forest reserves, Indian lands, or other locations to repeat the destructive process. Deforestation has increased exponentially in northwestern Brazil, and it has threatened the survival of indigenous peoples and an untold number of species of plants and animals. The import of disease proved disastrous for Rondonia's local Indian population, which dropped from 35,000 in 1965 to 6,000 in 1991.[33]

If planners had taken into account the interconnected environmental, social, and other noneconomic aspects of the projects they funded, their decisions might have been different, and they might have put the money they controlled toward other projects. Multilateral development banks could improve their work significantly by recognizing the interconnections among all the elements of society: economics, environment, culture, population, history, politics, and many others.

The United States wields great power in the development banks because it is the largest funder. Except for Japan, which recently increased its financing, no other country comes close to the influence the United States has at the World Bank, the Inter-American Development Bank, and similar institutions.

Because of the nature of international development projects, U.S. policy regarding those banks is inherently an environmental policy, a social policy, a population-related policy, and a foreign policy, all at once. Yet the United States has failed to include environmental, social, and other related experts in the development of its multilateral bank policies. The Treasury Department and the White House, along with the State Department, have been in control. The Environmental Protection Agency, the Department of Health and Human Services, the Department of Transportation, and others, have not been involved.

In international lending, as in all other efforts described in this book, policymakers need to understand the cycles that bind together our economic, environmental, social, and cultural systems. The health of the whole system depends on all the parts, and unless we acknowledge the patterns that connect these parts, our efforts to provide security and a better life for all will fail.

Afterword: Patterns & Principles

This book has attempted to point out the connections among our economic, environmental, social, and cultural systems. Although the patterns that appear in natural systems have their parallels in human activities, we often ignore them—and pay the price. To recognize and understand them at the local, national, and global level, it would be useful to apply a few basic principles that underlie the successes described here: inclusion of all that is affected by development, and of all that affects it; transparency of action; balance; and sustainability.

The principle of inclusion must be applied in many forms: inclusion of environmental and social developments in national accounting techniques; inclusion of local peoples in the planning of development projects, like those of the World Bank; inclusion of environmental and social assessments in national policy decisions; and inclusion of the largest numbers of people possible in political and economic systems.

Although the effect of such inclusion might be different in each case, the recipe followed for an increasingly genuine and deeply rooted system to support human development has the same bases. It is by taking advantage of human diversity and imagination and participation that the most well grounded solutions are found.

Human rights seek to promote this. At their fullest, they would guarantee participation in almost all aspects of society. As such, human rights and the principle of inclusion go hand in hand. Paralleling this is the principle of transparency. The openness that it calls for in government and other bodies enables people to determine where their actions and participation can have the most effect. It is at the source of nondiscrimination and interaction.

What about participation of nonhuman resources, like those of the environment? They too benefit from the principle of balance. In agriculture, for example, a pattern of farming that balances various plant and tree crops will control pests naturally, without extensive applications of chemicals that otherwise distort the balance. The same is true in institutions. At the United Nations, a great many agencies stand ready to provide their diverse perspectives and act as resources in balanced decision making. But when it comes to development planning, two U.N. agencies, the World Bank and the International Monetary Fund, often dominate decisions. A lack of balance, with economic models overwhelming environmental and social concerns such as health and education, can result, and it mirrors the dominance of chemical pesticides over balanced, natural methods of agriculture.

In virtually every part of this book, the principle of balancing future needs and hopes with those of today is represented. Looking ahead, to the future, is indispensable. Similarly, the principle of looking sideways is at the heart of most environmental, social, and economic ideas. It is the idea of interconnectedness. Weather patterns, rivers, and ocean currents tie regions together. Global warming, ozone depletion, and acid rain link all parts of the world. And what each one does today affects the potential that all others will have in the future. The future and the interconnected present are inseparable.

Actions that take advantage of cycles expand the potential of the future: recycling materials, reusing them, and reducing the amount of goods that ever enter the cycle of production. Actions that follow linear patterns of extraction of raw materials, production, transportation, consumption, and disposal only take from the earth; they do little or nothing to protect or expand options in the future.

Perhaps the principle that tries to pull all others together is that of sustainability. Sustainable development is a process of looking to the future, of understanding the cycles of nature and society, of expanding inclusion of people and perspectives, of conserving balance, and of accepting interconnectedness. It says that those are the elements of a future that will maximize rather than limit human potential.

To translate these principles into an effective effort to build an equitable and sustainable society in the United States, national leadership is essential, and a coherent framework for national action. But we must not wait for that to happen before working for change at the local level. As you decide on issues that you want to work on in your area, consider the patterns that connect the economic, environmental, social, and cultural systems there. And urge others—your representatives, local industry, public interest groups, teachers and students—to apply the principle of sustainable development, in all its meanings, to each new undertaking. You are the leaders who will connect with others to bring equity into development and to restore the environment. The solutions are in your programs, and the power to transform current practices will come from your communicating ever more effectively locally and globally.

Notes

CHAPTER 1

1. World Commission on Environment and Development, *Our Common Future* (Oxford: Oxford University Press, 1987), p. 43.

CHAPTER 2

1. Will Nixon, "The Greening of the Big Apple," *E Magazine*, September/October 1991, pp. 36-37.
2. Ibid.
3. *Ecocity Conference Report,* 1990, Urban Ecology, Berkeley, CA.
4. "Rating America's Cities," *E Magazine,* September/October 1991, pp. 38-39.
5. Rick Piltz and Sheila Machado, *Searching for Success* (Washington, D.C.: Renew America, 1990), p. 61.
6. Ibid.
7. Susan Meeker-Lowry, *Economics as If the Earth Really Mattered* (Philadelphia: New Society Publishers, 1988).
8. Huey D. Johnson, "Investing For Prosperity," Resource Renewal Institute, San Francisco, CA, August 1991, p. 1.
9. Keith Schneider, "Small Farms Sell Shares in a Way of Life," *New York Times,* July 1, 1990.
10. "CHOICE in Youngstown, Ohio," and "Northern Communities CLT, Duluth, Minnesota," both in *Community Economics,* Spring 1991, pp. 8-11, 14.
11. Meeker-Lowry, *Economics as If the Earth Really Mattered,* pp. 105-06.
12. Bob Hall and Mary Lee Kerr, *1991-1992 Green Index* (Washington, D.C.: Island Press, 1991).

13. James J. MacKenzie, "Toward A Sustainable Energy Future: The Critical Role of Rational Energy Pricing," World Resources Institute, Washington, D.C., May 1991, p. 12.
14. Ibid., pp. 6, 7.
15. Concern, Inc., *Global Warming & Energy Choices* (Washington, D.C.: 1991), p. 28.
16. Ibid.
17. Ibid., p. 29.
18. Ibid., p. 26.
19. Ibid., pp. 26-27.
20. MacKenzie, "Toward a Sustainable Energy Future," p. 2.
21. Ibid.
22. Ibid.
23. "National Energy Strategy, Executive Summary," First Edition 1991/1992, Washington, D.C., February 1991, p. 17.
24. MacKenzie, "Toward a Sustainable Energy Future," p. 2.
25. U.S. Citizens Network on U.N. Conference on Environment and Development, Working Group Report on Energy, "A Sustainable Energy Policy for the United States," p. 3.
26. UCS recommendations contained in *Nucleus,* Spring 1991, as reprinted in Citizens Network, Working Group Report on Energy.
27. Johnson, "Investing For Prosperity," p. 2.
28. Marc H. Ross and Robert H. Socolow, "Synthesis of the Symposium Session," in World Resources Institute, *Technological Trans formation for Sustainable Development* (Washington, D.C.: January 1991), pp. 12-15.
29. Ibid.
30. Larry Martin, "Hazardous Waste Minimization Technology Transfer," Hazardous and Solid Waste Minimization, Government Institutes, Inc., p. 9-1.
31. Lester R. Brown et al., *Saving the Planet* (New York: W.W. Norton & Co., 1991), p. 144.
32. Ibid., p. 143.
33. Ibid.
34. Ibid., p. 144.
35. Ibid., pp. 147-48.
36. Ibid., p. 145.

CHAPTER 3

1. Testimony of Richard J. Smith, Head of U.S. Delegation, Sofia CSCE Meeting on the Protection of the Environment, Before The Commission on Security and Cooperation in Europe, September 28, 1989, pp. 5, 6.
2. Ibid.
3. Declaration of the United Nations Conference on the Human Environment, Stockholm, Sweden, June 16, 1972, Principles 1 and 2, from Michael Kane, *International Protection of Human Rights and the Environment,* U.S. Environmental Protection Agency, May 1991, unpublished, p. 46.
4. Walter S. Mossberg, "Renewed Superpower Interest Pumps Life Into Helsinki Negotiating System," *Wall Street Journal,* December 4, 1989.
5. Ibid.
6. Ibid.
7. Organisation for Economic Co-operation and Development, "Joint Report on Trade and Environment," Paris, April 29, 1991, p. 1.
8. The United Nations Association of the United States of America (UNA-USA) and the Sierra Club, *Uniting Nations for the Earth,* A Report of the Multilateral Project (New York: UNA-USA, 1990).
9. International Chamber of Commerce, Second World Conference on Environmental Management, Conference Report and Back ground Papers (the conference), April 10-12, 1991, Rotterdam, p. 26.
10. Ibid., p. 28.
11. "Whatever Happened With the Valdez Principles?" *Business Ethics,* May/June 1991, p. 11.
12. Larry Reynolds, "Vice President of the Environment," *Business Ethics,* March/April 1991, pp. 22-24.
13. "Pulse of the People," *The Green Consumer Letter,* January 1991, p. 2.
14. "Power to the People," *The Green Consumer Letter,* February 1991, p. 2.
15. Lester R. Brown et al., *Saving the Planet* (New York: W.W. Norton & Co., 1991), p. 152.
16. Walter V. Reid et al., *Bankrolling Successes: A Portfolio of Sustain able Development Projects* (Washington, D.C.: Environmental Policy Institute and National Wildlife Federation, 1988), pp. 6-7.

17. Ibid.
18. Ibid., pp. 26-27.
19. Ibid.
20. Memorandum from Steve Romanoff to Andy Maguire on ATI's Environmental Activities and Possible Initiatives, Appropriate Technology International, Washington, D.C., September 5, 1990, p. 9.
21. Chet Atkins, "International Family Planning: Where's the Leader ship?" *Washington Post,* August 27, 1991.
22. Jodi L. Jacobson, Worldwatch Institute, based on data from Population Crisis Committee, private communication to Linda Starke, September 13, 1991.
23. Reid et al., *Bankrolling Successes,* pp. 32-33.
24. Keith Bradsher, "U.S. Ban on Mexico Tuna Is Overruled," *New York Times,* August 23, 1991.

CHAPTER 4

1. Herman E. Daly and John B. Cobb, Jr., *For the Common Good* (Boston: Beacon Press, 1989), p. 2.
2. U.S. Citizens Network on the UN Conference on Environment and Development, Working Group Report on Economics, Financial Institutions, and Militarism, "Faulty Premises Underlying Development Theory Used by International Financial Institutions," p. 1.
3. Office of Technology Assessment, U.S. Congress, *Serious Reduction of Hazardous Waste* (Washington, D.C.: U.S. Government Printing Office, 1986).
4. *Air/Water Pollution Report,* Vol. 29, No. 37.
5. "EPA Proposes Monitoring Municipal Dumps," *Washington Post,* August 25, 1988.
6. Robert Repetto et al, *Wasting Assets: Natural Resources in the National Income Accounts* (Washington, D.C.: World Resources Institute, 1989), p. 2.
7. Ibid., p. 4.
8. Quoted in Linda Starke, *Signs of Hope* (Oxford: Oxford University Press, 1990), p. 140.
9. Lester R. Brown et al., *Saving the Planet* (New York: W.W. Norton & Co., 1991), p. 125.
10. Repetto et al., *Wasting Assets,* pp. 9-11.

11. Ibid., p. 11.
12. U.N. Development Programme (UNDP), *Human Development Report 1991* (New York: Oxford University Press, 1991).
13. Ibid., pp. 17, 119.
14. Revised ISEW figures from Brown et al., *Saving the Planet,* p. 127.
15. Keith Schneider, "Science Academy Says Chemicals Do Not Necessarily Increase Crops," *New York Times,* September 8, 1989.
16. UNDP, *Human Development Report 1991,* p. 80.
17. World Resources Institute (WRI) and International Institute for Environment and Development (IIED), *World Resources 1988-89* (New York: Basic Books, Inc., 1988), p. 204.
18. Brown et al., *Saving the Planet,* p. 132.
19. WRI and IIED, *World Resources 1988-89,* p. 212.
20. Brown et al., *Saving the Planet,* 133.
21. Ibid., pp. 134-35.
22. Ibid., p. 135.
23. Global Tomorrow Coalition (GTC), *Global Ecology Handbook* (Boston: Beacon Press, 1990), p. 44; WRI and IIED, *World Resources 1988-89,* p. 242.
24. Susan George, "Third World Debt: The Moral and Physical Equivalent of War," *Who Owes Whom?,* Newsletter of Project Abraco: North Americans in Solidarity with the People of Brazil, Spring 1988.
25. Ruth Leger Sivard, *World Military and Social Expenditures 1991* (Washington, D.C.: World Priorities, 1991), p. 48.
26. GTC, *Global Ecology Handbook,* p. 44; WRI and IIED, *World Resources 1988-89,* p. 242.
27. Walden Bello and Stephanie Rosenfeld, *Dragons in Distress: Asia's Miracle Economies in Crisis* (San Francisco, CA: Institute for Food and Development Policy, 1990), pp. 37, 183.
28. Global Tomorrow Coalition and The Centre for Our Common Future, *Sustainable Development: A Guide to Our Common Future* (Washington, D.C.: 1989), pp. 23-24.
29. Sivard, *World Military and Social Expenditures 1991,* p. 7.
30. Dr. Harald Müller, Director, Frankfurt Peace Research Institute, and Adjunct Professor of Security Studies, Bologna Center of the School of Advanced International Studies, Johns Hopkins University, private communication, February 1991.

31. Sivard, *World Military and Social Expenditures,* pp. 27, 11.
32. Ibid., p. 46.
33. Ibid., p. 53.
34. World Bank, *World Development Report 1991* (New York: Oxford University Press, 1991), pp. 205, 257, 259, 266-67.
35. UNDP, *Human Development Report 1991,* p. 119.
36. Ibid., p. 81.
37. Brown et al., *Saving the Planet,* p. 151.

CHAPTER 5

1. Walter V. Reid and Kenton R. Miller, *Keeping Options Alive: The Scientific Basis for Conserving Biodiversity* (Washington, D.C.: World Resources Institute, 1989), pp. 25, 3.
2. Ibid., p. 28.
3. Jeffrey A. McNeely et al., *Conserving the World's Biological Diversity* (Gland, Switzerland, and Washington, D.C.: International Union for Conservation of Nature and Natural Resources, World Resources Institute, Conservation International, World Wildlife Fund-US, and World Bank, 1990), p. 9.
4. Reid and Miller, *Keeping Options Alive,* p. 33.
5. William K. Stevens, "Species Loss: Crisis or False Alarm?" *New York Times,* August 20, 1991.
6. Reid and Miller, *Keeping Options Alive,* p. 49.
7. World Resources Institute (WRI), *World Resources 1990-91* (New York: Oxford University Press, 1990), p. 126.
8. Reid and Miller, *Keeping Options Alive,* pp. 40-41.
9. Global Tomorrow Coalition (GTC), *Global Ecology Handbook* (Boston: Beacon Press, 1990), p. 118.
10. Lester R. Brown et al, *Saving the Planet* (New York: W.W. Norton & Co., 1991), p. 76.
11. Lester R. Brown et al., *State of the World 1991* (New York: W.W. Norton & Co., 1991), p. 75.
12. Consortium for Action to Protect the Earth, *Forests* (draft), Washington, D.C., June 13, 1991, unpublished, p. 2.
13. WRI, *World Resources 1990-91,* p. 103.
14. Brown et al., *State of the World 1991,* p. 75.
15. Robert Repetto, "Deforestation in the Tropics," *Scientific American,* April 1990, p. 36.

16. Brown et al., *State of the World 1991,* p. 75.
17. Julian Burger, *The Gaia Atlas of First Peoples* (New York: Anchor Books, 1990), p. 88.
18. U.S. Citizens Network on the U.N. Conference on Environment and Development, Working Group Report on Oceans, p. 1.
19. GTC, *Global Ecology Handbook,* p. 138.
20. Citizens Network, Working Group Report on Oceans, pp. 4, 5.
21. GTC, *Global Ecology Handbook,* p. 139.
22. Office of Technology Assessment, U.S. Congress, *Serious Reduction of Hazardous Waste* (Washington, D.C.: U.S. Government Printing Office, 1986), p. 18.
23. Citizens Network, Working Group Report on Oceans, p. 4.
24. Ibid.
25. Chesapeake Bay Program Executive Council, "Progress Report 1990-1991," U.S. Environmental Protection Agency Region III, Annapolis, MD.
26. Lawrence Ingrassia, "Overfishing Threatens to Wipe Out Species and Crush Industry," *Wall Street Journal,* July 16, 1991.
27. WRI, *World Resources 1990-91,* p. 180.
28. GTC, *Global Ecology Handbook,* p. 144.
29. Citizens Network, Working Group Report on Freshwater Resources, pp. 4-5.
30. GTC, *Global Ecology Handbook,* p. 158.
31. Quoted in Citizens Network, Working Group Report on Freshwater Resources, p. 3.
32. Ibid., p. 4.
33. GTC, *Global Ecology Handbook,* p. 159.
34. Ibid.
35. Ibid., pp. 163, 164.
36. Lester R. Brown et al., *State of the World 1990* (New York: W.W. Norton & Co., 1990), p. 54.
37. Sandra Postel, "California's Liquid Deficit," *New York Times,* February 27, 1991.
38. GTC, *Global Ecology Handbook,* p. 160.
39. WRI, *World Resources 1990-91,* p. 166.
40. Brown et al., *Saving the Planet,* p. 84.
41. Brown et al., *State of the World 1990,* p. 59.
42. Peter Weber, Worldwatch Institute, Washington, D.C., private communication to Linda Starke, August 28, 1991.

43. GTC, *Global Ecology Handbook,* p. 80.
44. Ibid., p. 84.
45. Keith Schneider, "Science Academy Says Chemicals Do Not Necessarily Increase Crops," *New York Times,* September 8, 1989.
46. WRI, *World Resources 1990-91,* p. 98.
47. Schneider, "Science Academy Says Chemicals Do Not Necessarily Increase Crops."
48. GTC, *Global Ecology Handbook,* p. 247.
49. Ibid., p. 249.
50. Heather Spalding, Waste Trade Office, Greenpeace, Washington, D.C., private communication to Linda Starke, August 22, 1991.
51. GTC, *Global Ecology Handbook,* p. 250.
52. Global Tomorrow Coalition and The Centre for Our Common Future, *Sustainable Development: A Guide to Our Common Future* (Washington, D.C.: 1989), p. 45.
53. Todd quoted in Donella Meadows, "The New Alchemist," *Harrowsmith Magazine,* November-December 1988, pp. 38-39.
54. Abyd Karmali, "Stimulating Cleaner Technologies Through the Design of Pollution Prevention Policies: An Analysis of Impedi ments and Incentive," Master's Thesis, Technology and Policy Program, Massachusetts Institute of Technology, 1990.
55. Lester R. Brown et al., *State of the World 1988* (New York: W.W. Norton & Co., 1988), p. 122.
56. GTC, *Global Ecology Handbook,* p. 253.
57. Anders Wijkman and Lloyd Timberlake, *Natural Disasters: Acts of God or Acts of Man?* (London and Washington, D.C.: International Institute for Environment and Development, 1984), p. 11.
58. Lloyd Timberlake, *Africa In Crisis* (London and Washington, D.C.: International Institute for Environment and Development, 1985), p. 19.
59. Ibid., p. 21.
60. Kenneth Brower et al., *One Earth* (San Francisco, CA: Collins Publishers, 1990), p. 124.
61. Ibid., p. 19.
62. Wijkman and Timberlake, *Natural Disasters,* p. 11.
63. Merbold quoted in Brower et al., *One Earth,* p. 75.
64. Brown et al., *State of the World 1990,* p. 98.
65. GTC, *Global Ecology Handbook,* p. 223.
66. Ibid.

67. WRI, *World Resources 1990-91,* pp. 201-02.
68. Brown et al., *State of the World 1990,* p. 99.
69. Ibid., p. 98.
70. WRI, *World Resources 1990-91,* p. 202.
71. William K. Stevens, "Ozone Loss Over U.S. Is Found To Be Twice as Bad as Predicted," *New York Times,* April 5, 1991.
72. Citizens Network, Working Group Report on Energy for a Sustainable World, "The Stratospheric Ozone Depletion Crisis," p. 1.
73. Stevens, "Ozone Loss Over U.S."
74. Colin Hines, Greenpeace UK, London, private communication to Linda Starke, August 2, 1991.
75. Citizens Network, "Stratospheric Ozone Depletion Crisis," p. 1.
76 WRI, *World Resources 1990-91,* p. 13.
77. Ibid., p. 15.
78. Citizens Network, Working Group Report on Energy for a Sustainable World, "A Sustainable Energy Policy for the United States," p. 2.
79. Brown et al., *Saving the Planet,* p. 36.
80. "No Agreement in Geneva," *Atmosphere* (Friends of the Earth International), February 1991, pp. 1, 6.
81. Citizens Network, "A Sustainable Energy Policy for the United States," p. 1.
82. Ibid.

CHAPTER 6

1. U.N. Development Programme (UNDP), *Human Development Report 1991* (New York: Oxford University Press, 1991), pp. 142-43, 174.
2. Ibid., p. 137.
3. Jodi L. Jacobson, *Women's Reproductive Health: The Silent Emergency* (Washington, D.C.: Worldwatch Institute, 1991), pp. 5, 12.
4. United Nations Children's Fund (UNICEF), *The State of the World's Children 1990* (New York: Oxford University Press, 1990), p. 17.
5. UNDP, *Human Development Report 1991,* p. 141.
6. Bob Hall and Mary Lee Kerr, *1991-1992 Green Index* (Washington, D.C.: Island Press, 1991), p. 83.

7. Ibid.
8. Ibid.
9. World Bank, *World Development Report 1991* (New York: Oxford University Press, 1991), p. 259.
10. Global Tomorrow Coalition and The Centre for Our Common Future, *Sustainable Development: A Guide to Our Common Future* (Washington, D.C.: 1989), p. 45.
11. Kevin P. Phillips, "Reagan's America: A Capital Offense," *New York Times Magazine,* June 17, 1990, p. 26; U.S. Citizens Network on the UN Conference on Environment and Development, Working Group Report on Economics, Financial Institutions, and Militarism, "Poverty in the United States," p. 4.
12. Phillips, "Reagan's America," p. 26.
13. Jason DeParle, "Poverty Rate Rose Sharply Last Year as Incomes Slipped," *New York Times,* September 27, 1991.
14. UNDP, *Human Development Report 1991,* p. 31; David Whitman, "The Rise of the "Hyper-Poor"," *U.S. News & World Report,* October 15, 1990, p. 40.
15. Jason DeParle, "In Rising Debate on Poverty, the Question: Who Is Poor?" *New York Times,* September 3, 1990; Spencer Rich, "Economist Says Poverty Line Should Be Higher," *Washington Post,* May 4, 1990.
16. UNDP, *Human Development Report 1991,* p. 31.
17. Timothy M. Smeeding, "Children and Poverty: How U.S. Stands," *Forum for Applied Research and Public Policy,* Summer 1990, p. 65.
18. Lester R. Brown et al., *State of the World 1991* (New York: W.W. Norton & Co., 1991), p. 90.
19. World Resources Institute, *World Resources 1990-91* (New York: Oxford University Press, 1990), p. 92.
20. "Why Women?" (brochure), Women's Environment & Development Organization, Women USA Fund, Inc., New York, undated.
21. UNDP, *Human Development Report 1991,* p. 31.
22. "Women and Children First" Final Report (draft), United Nations Conference on Environment & Development, Conches, Switzerland, June 11, 1991, p. 16.
23. Citizens Network, "Poverty in the United States," p. 6.
24. Julian Burger, *The Gaia Atlas of First Peoples* (New York: Anchor Books, 1990), p. 18.
25. Ibid., pp. 24, 30, 32.

26. Ibid., p. 122.
27. Ibid., p. 181.
28. The Sierra Club, *Bankrolling Disasters* (Washington, D.C.: 1986), p. 6.
29. Ibid.
30. Ibid.
31. Ibid., p. 5.
32. James Brooke, "Plan to Develop Amazon a Failure," *New York Times,* August 25, 1991.
33. Ibid.

Index

About the Authors

HAL KANE works in Washington, D.C., where he writes on environment and development issues. He attended the University of Michigan and the Johns Hopkins University of Advanced International Studies, for undergraduate and graduate school, respectively. He has organized venture capital conferences and has worked on various documents and speeches for several nonprofit organizations and the Environmental Protection Agency. Currently he is focusing on trade policy and other connections between economics and the environment.

LINDA STARKE has been working on population and environment issues for more than twenty years, and has been a freelance writer and editor since 1982. She is the author of *Signs of Hope: Working Towards Our Common Future,* a summary of developments since the 1987 report of the World Commission on Environment and Development, which she edited. She edits and oversees the production of *State of the World*, Worldwatch Institute's annual review of progress toward a sustainable society. She has also worked with the Office of Technology Assessment, the U.N. Development Programme, the World Bank, the World Conservation Union, and the World Resources Institute. Her articles have appeared in the *New York Times*, the *Washington Post*, and *Business Ethics* magazine.

Also Available From Island Press

Ancient Forests of the Pacific Northwest
By Elliott A. Norse

Balancing on the Brink of Extinction: The Endangered Species Act and Lessons for the Future
Edited by Kathryn A. Kohm

Better Trout Habitat: A Guide to Stream Restoration and Management
By Christopher J. Hunter

Beyond 40 Percent: Record-Setting Recycling and Composting Programs
The Institute for Local Self-Reliance

The Challenge of Global Warming
Edited by Dean Edwin Abrahamson

Coastal Alert: Ecosystems, Energy, and Offshore Oil Drilling
By Dwight Holing

The Complete Guide to Environmental Careers
The CEIP Fund

Economics of Protected Areas
By John A. Dixon and Paul B. Sherman

Environmental Agenda for the Future
Edited by Robert Cahn

Environmental Disputes: Community Involvement in Conflict Resolution
By James E. Crowfoot and Julia M. Wondolleck

Forests and Forestry in China: Changing Patterns of Resource Development
By S. D. Richardson

The Global Citizen
By Donella Meadows

Hazardous Waste from Small Quantity Generators
By Seymour I. Schwartz and Wendy B. Pratt

Holistic Resource Management Workbook
By Allan Savory

In Praise of Nature
Edited and with essays by Stephanie Mills

The Living Ocean: Understanding and Protecting Marine Biodiversity
By Boyce Thorne-Miller and John G. Catena

Natural Resources for the 21st Century
Edited by R. Neil Sampson and Dwight Hair

The New York Environment Book
By Eric A. Goldstein and Mark A. Izeman

Overtapped Oasis: Reform or Revolution for Western Water
By Marc Reisner and Sarah Bates

Permaculture: A Practical Guide for a Sustainable Future
By Bill Mollison

Plastics: America's Packaging Dilemma
By Nancy Wolf and Ellen Feldman

The Poisoned Well: New Strategies for Groundwater Protection
Edited by Eric Jorgensen

Race to Save the Tropics: Ecology and Economics for a Sustainable Future
Edited by Robert Goodland

Recycling and Incineration: Evaluating the Choices
By Richard A. Denison and John Ruston

Reforming the Forest Service
By Randal O'Toole

The Rising Tide: Global Warming and World Sea Levels
By Lynne T. Edgerton

Saving the Tropical Forests
By Judith Gradwohl and Russell Greenberg

Trees, Why Do You Wait?
By Richard Critchfield

War on Waste: Can America Win Its Battle With Garbage?
By Louis Blumberg and Robert Gottlieb

Western Water Made Simple
From *High Country News*

Wetland Creation and Restoration: The Status of the Science
Edited by Mary E. Kentula and Jon A. Kusler

Wildlife and Habitats in Managed Landscapes
Edited by Jon E. Rodiek and Eric G. Bolen